Cleaning-up the Ganges

Cleaning-up the Ganges
A Cost–Benefit Analysis of the
Ganga Action Plan

A. Markandya

M.N. Murty

OXFORD
UNIVERSITY PRESS

OXFORD
UNIVERSITY PRESS

YMCA Library Building, Jai Singh Road, New Delhi 110001

Oxford University Press is a department of the University of Oxford. It
furthers the University's objective of excellence in research, scholarship, and
education
by publishing worldwide in

Oxford   New York

Athens   Auckland   Bangkok Bogota Buenos Aires Calcutta
Cape Town   Chennai   Dar es Salaam   Delhi   Florence   Hong Kong   Istanbul
Karachi   Kuala Lumpur   Madrid   Melbourne   Mexico City   Mumbai
Nairobi Paris Sao Paolo Shanghai Singapore Taipei Tokyo Toronto
Warsaw

with associated companies in Berlin Ibadan
Oxford is a registered trade mark of Oxford University Press
in the UK and in certain other countries

Published in India
By Oxford University Press, New Delhi

ISBN 019 564945 1

Typeset by Urvashi Press, Meerut, UP
Printed at Roopak Printers, Delhi 110 032
Published by Manzar Khan, Oxford University Press
YMCA Library Building, Jai Singh Road, New Delhi 110 001

Preface

The material presented in this book is based on the research project 'Cost Benefit Analysis of Ganga Action Plan', funded by the Overseas Development Administration (ODA), UK through the Ministry of Environment and Forests, Government of India. The major part of the project was undertaken by the University of Bath and Metroeconomica, UK and the Institute of Economic Growth, Delhi. The All India Institute of Hygiene and Public Health (AIH&PH), Calcutta, the Industrial Toxico-logy Research Centre (ITRC), Lucknow, and the Inland Capture Fisheries Research Institute, West Bengal have contributed respectively to Chapters 7 (the Health Impacts of the GAP), Chapter 8 (the impact of GAP projects on Wastewater Toxicants), and Chapter 9 (Fisheries on the River Ganga). Chapter 11 on the biodiversity of the Ganges is based on the report prepared by Dr P. Sinha of Patna University for the project.

Apart from ourselves, many people from Metroeconomica, UK and the Institute of Economic Growth, India have participated in the project. In particular, we acknowledge with thanks the contributions of Mr T. Taylor of Bath University, Dr D. Duthie of Metroeconomica, UK, and Dr A.J. James, Dr U.R. Prasad, Mrs Smita Misra, and Mr S.S. Yadav, all at the Institute of Economic Growth.

The work at AIH&PH was carried out by Professor S. Nath and his colleagues while the work at ITRC was undertaken by Dr K. Singh. We wholeheartedly thank them for their valuable contribution to the project.

Eco-Tech Services Limited, Delhi, undertook the job of coordinating the project work. We express our thanks to the late Mr Arun Kumar, Ms Jyoti Mathur, and Ms Sonali Pachauri for efficiently coordinating the project activities and organizing a number of workshops and the final seminar.

We are grateful to Mrs Malti Sinha and Mrs Pushpa Thottan of the Ministry of Environment and Forest, Government of India for their co-operation and help in executing this project.

We also express our gratitude to Mr Ian Curtis, Mr Brian Baxendale, and their colleagues at the Department for International Development (DFID), Delhi for the encouragement and financial support for this project.

It is obvious that a study like the cost–benefit analysis of cleaning a major water body like the Ganges should have a multi-disciplinary approach. Given the complexity of the problem, such a study has to be jointly undertaken by economists, ecologists, scientists, and health experts. This book is the first of its kind in India and probably one of the few studies made abroad using a multi-disciplinary approach. Given the constraints on time and resources, only certain aspects of the problem are dealt with in detail while others require further in-depth study, for their completeness. We believe that this book while highlighting the difficulties involved, nevertheless provides insights into the role of different disciplines of academics in the measurement of benefits from the improved water quality of a major river like the Ganges.

Finally, we wish to mention our debt to Ms Nitasha Devasar and Dr Kavita Iyengar of Oxford University Press, New Delhi for the interest taken in the original manuscript and encouragement for the publication of this book.

July 2000   A. MARKANDYA

                                                               M.N. MURTY

Contents

Figures

Tables

Abbreviations and Acronyms

ADI	acceptable daily intake
AIIH&PH	All India Institute of Hygiene and Public Health
AOX	Absorbable Organic Halides
BAT	best available technology
BEP	best environmental practice
BHC	benzenehexachloride
BOD	biological oxygen demand
Ca	calcium
Cd	cadmium
CEC	Central European Countries
c.i.f.	cost, insurance and freight
CIFRI	Central Inland Capture Fisheries Research Institute
CITES	Convention on Trade in Endangered Species
COD	chemical oxygen demand
CPCB	Central Pollution Control Board
Cr	chromium
Cu	copper
CUPE	catch per unit of effort
CV	Contingent Valuation
CVM	Contingent Valuation Method
CWC	Central Water Commission
d	day
DALYs	disability adjusted life years
DAP	diammonium phosphate
DDT	dichlorodiphenyle-trichloroethane
DFID	Department for International Development
DO	dissolved oxygen
DRF	dose-response function

EC	European Commission
EOCL	Extractable Organic Chlorine
EPA	Environmental Protection Agency
EQO	Environmental Quality Objectives
EQOs	environmental quality objectives
ETP	effluent treatment plant
EU	European Union
Fe	Iron
f.o.b.	free on board
GAP	Ganga Action Plan
GEF	Global Environment Facility
HCB	Hexachlorobenzene
HCH	Hexachlorocyclohexane
ICPR	International Commission for the Protection of the Rhine against Pollution
IEG	Institute of Economic Growth
IRC	International River Commission
IRR	internal rate of return
ITRC	Industrial Toxicology Research Centre
IUCN	International Union for Conservation of Nature and Natural Resources
K	Potassium
KL	kilolitre
l	litre
Mg	magnesium
MLD	million litres per day
Mn	Manganese
MoEF	Ministry of Environment and Forests
MPPCB	Madhya Pradesh Pollution Control Board
MPS	Municipal Pumping Station
N	nitrogen
Na	sodium
NAP	National Action Plan
NEERI	National Environmental Engineering Research Institute
NGO	non-governmental organization

Ni	nickel
NOAA	National Oceanic and Atmospheric Administration
NPV	net present value
NRCD	National River Conservation Directorate
ODA	Overseas Development Administration
OECD	Organization for Economic Cooperation and Development
OLS	ordinary least squares
Org C	organic carbon
P	phosphorus
PAHs	Polycyclic Aromatic Hydrocarbons
Pb	lead
PCBs	Polychlorinated biphenyls
PCU	Programme Coordinating Unit
pH	Acidity
PS	pumping station
RAP	Rhine Action Programme
RQ	risk quotient
SAP	Strategic Action Plan for the Danube River Basin
SMU	social marginal utility
STPs	Sewerage Treatment Plants
TDI	total daily intake
TEV	Total Economic Value
UK	United Kingdom
UNDP	United Nations Development Program
UNEP	United Nations Environmental Program
UNIDO	United Nations Industrial Development Organization
UP	Uttar Pradesh
VOSL	valuation of a statistical life
WHO	World Health Organization
WQM	Water Quality Modelling
WTA	willingness to accept
WTP	willingness to pay
Zn	zinc

CURRENCIES

DM	Deutsche Mark
£	Pound Sterling
US$	US Dollar
$	Dollar
Rs	Rupees
Dgl	Dutch guilders
ECU	European Currency Unit
£ (Egyptian)	

1

Introduction

The Ganga is the most important river system in India and one of the most important in the world. It has a basin covering 861,404 square kilometres. Already half a billion people—almost one-tenth of the world's population—live within the river basin at an average density of over 500 per square kilometre, and this population is projected to increase to over one billion people by the year 2030.

The Ganges plains were first settled by Aryan tribes around 1200 BC and, in the subsequent 3200 years of occupation, the landscape of the region has been completely transformed by over one hundred generations of agriculturists and the recent expansion of industrial activity.

Today the 2510 km long river supports 29 Class I cities, 23 Class II cities and 48 towns, plus thousands of villages. Nearly all the sewage from these populations goes directly into the river, totalling over 1.3 billion litres per day, along with a further 260 million litres of industrial waste, run-off from the 6 million tons of fertilizers and 9000 tons of pesticides used in agriculture within the basin, and large quantities of solid waste, including thousands of animal carcasses and several hundred human corpses released into the river every day for spiritual rebirth. Figure 1.1 provides an illustrative map of the Ganges river. A schematic map of the river is provided in Figure 1.2, indicating water quality in terms of dissolved oxygen in 1985–6. This is only intended to provide a reference set of conditions and a guide to locations; it is not an accurate representation of the basin.

The inevitable result of this onslaught on the river's capacity to receive and assimilate waste has been an erosion of river water quality, to the extent that by the 1970s, large stretches (over 600 kilometres) of the river were effectively dead from an ecological point of view and posed a considerable public health threat to the thousands of religious bathers using the river every day.

The 'Ganga Action Plan (GAP)' originated from the personal intervention and interest of the late Prime Minister Indira Gandhi, who

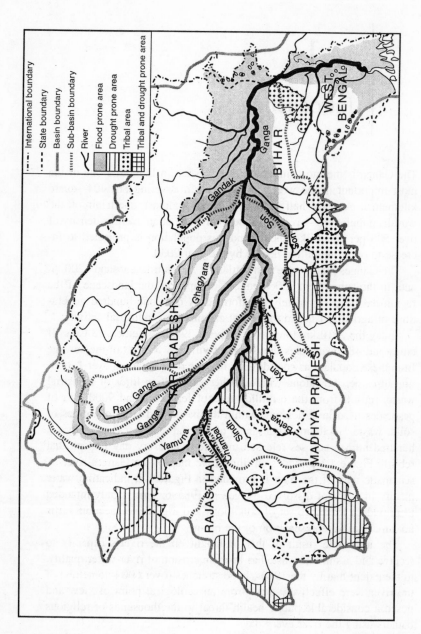

FIGURE 1.1: An Illustrative Map of the River Ganga

Legend:
- International boundary
- State boundary
- Basin boundary
- Sub-basin boundary
- River
- Flood prone area
- Drought prone area
- Tribal area
- Tribal and drought prone area

requested a comprehensive survey of the situation in 1979. After five years, the Central Pollution Control Board (CPCB) published two comprehensive reports which formed the base from which the action plan to clean-up the Ganga was developed.

In February 1985, the Central Ganga Authority was established, with an initial budget allocation of around Rs 350 crore (Rs 3.5 billion) to administer the cleaning of the Ganga and to restore it to a pristine condition. Implementation of the first phase of the GAP was delayed by the assassination of Indira Gandhi, and it was not until February 1985 that the GAP was finally launched. On this occasion the late Prime Minister Rajiv Gandhi stated:

The Ganga is a symbol of our spirituality, our tradition, our tolerance and our synthesis. But it is the most polluted river with sewerage and pollution from cities and industries thrown into it. From now, we shall put a stop to all this. We are launching this plan—not for the Public Works Department, but for the people of India.

The final cost of the GAP has been estimated at Rs 700 crore or Rs 7 billion for phase I and Rs 420 crore or Rs 4.2 billion for phase II. The operating costs of the programme run to around Rs 356 million.[1]

The GAP has been perhaps the largest single attempt to clean up a polluted river anywhere in the world. Although a number of other international scale river basin clean-up programmes have been effectively implemented in other countries, none has the full spectrum of geographical, ecological, and socio-cultural complexities which faced the Indian government during the implementation of the GAP. The sums of money referred to above are large by any standards, and were committed with the main objective of raising the river water quality to bathing standard. No attempt was made to compare costs and benefits of the programme. This is not unusual in river clean-up programmes, as similar programmes in Europe have also been carried out without a cost–benefit study. However, the Government of India decided, in 1995, to carry out such a study, largely on an *ex post* basis. With financial support from the Department for International Development (DFID) of the UK, the study team was assembled and this research commenced.

The objective of the study was to provide a comprehensive analysis of the costs and benefits of the schemes within the GAP and to quantify,

---

[1]The average exchange rates in 1995 were Rs 54.5 to the Pound Sterling and Rs 35.2 to the US Dollar. Hence the investment costs can be expressed as £205 million or $318 million. The operating costs are £6.4 million or $10 million.

to the maximum extent possible, the monetary values attached to those costs and benefits. It has been a multi-disciplinary study involving specialists in water quality modelling, environmental economics, epidemiology, toxicology, agriculture, fisheries, and biodiversity.

Before discussing the structure of the book, it is worth asking what has happened to water quality in the river as a result of the GAP. Whatever valuation is placed on the changes, to many people it is the physical improvement in water quality that is the most important.

In absolute terms the quality of water in the Ganga has shown varying improvements since 1985. More details are provided in Chapter 10 (see Table 10.5), but essentially dissolved oxygen levels have been improving in the areas of Kanpur, Allahabad, and Varanasi after 1992. In the lower stretch, at Nawabganj, however, dissolved oxygen levels have continued to decline. Similar improvements in phosphate and nitrate concentrations have been observed since the early 1990s.

An assessment of river water quality changes as a result of the GAP, however, cannot be made from the above figures. What is needed, instead, is a comparison between what the conditions would have been in the late 1990s without the GAP and the conditions with the GAP. Such a comparison has been carried out as part of this study using a sophisticated water quality model. The details are reported in Appendix A. The results of the model (see Table A.1 and the following discussion) show that some improvements in water quality [measured in terms of dissolved oxygen and biochemical oxygen demand (BOD)] were observed everywhere, although the improvements were quite small in some places. It is also worth noting that a total stretch of about 437 kilometres still violates the maximum permissible level of 3.0 milligram per litre of BOD. In terms of dissolved oxygen, the level throughout the river is now more than 5.0 milligram per litre. Without the GAP, more than 740 kilometres would have violated the BOD limit, with about 1000 kilometres having BOD levels in excess of 10 milligram per litre.

One can conclude, then, that some improvements in water quality have been achieved. The important question is, what are these worth in money terms, taking account of the broadest set of values placed on cleaner water? It is this question that this book focuses on.

Chapter 2 begins by looking at river clean-up programmes in other countries to assess what lessons could be learned for the GAP. It examines the histories of clean-up programmes for the Thames, the Rhine, and the Danube and tries to draw out the lessons for the GAP.

Chapter 3 describes briefly the methodology used in the study, which is one of extended cost–benefit analysis. All costs and benefits are estimated—both those that have money flows associated with them and those that do not. The comparison of costs and benefits is over a period of time (in this case the years 1995 to 2034). Costs and benefits in future periods are discounted using a real rate of interest of 10 per cent. This rate is applied as the cut-off rate for public sector projects in India. Net benefits are computed in terms of market prices as well as in terms of 'social' prices. The latter will differ from the former when the market prices of inputs and outputs are not equal to the social values attached to the same inputs and outputs. The summary statistics for the project are the net present value (NPV) and the Internal Rate of Return (IRR). Both are given in Chapter 12.

Chapter 4 provides the data on the total costs of the projects. GAP projects are divided into those associated with GAP I, the capital expenditures of which have effectively been completed, and GAP II, which was started in 1993 and will continue for a few more years. The investment cost data are broken down into domestic materials and equipment, skilled labour, and unskilled labour. This breakdown is necessary to estimate the economic costs and different shadow prices that apply to the different categories of costs. The same categories apply to the operating costs.

Chapters 5 to 9 provide an analysis of the benefits. Chapter 5 begins by looking at the non-user benefits. These are described as the benefits to those who do not visit the Ganga but have a willingness to pay to know that the river is purer and cleaner. The method used for the elicitation of these benefits is the contingent valuation method (CVM), described in some detail in Chapter 5. It consists of a carefully designed set of questionnaires to survey sample households in eleven major cities in India, which evoked unexpectedly good responses about the households' valuation of non-user benefits for an important water resource in the country. This work was undertaken by the Institute of Economic Growth (IEG) of Delhi University, working with Metroeconomica, a UK consulting firm.

Chapter 6 reports on the CVM survey results for user benefits. A carefully designed questionnaire was employed to survey a sample of urban households living near the banks of the river in the cities of Calcutta, Varanasi, and Allahabad. It attempted to measure only user benefits accruing to urban populations living near the Ganga as it flows through the three states of Uttar Pradesh, Bihar, and West Bengal. There can obviously be substantial user benefits accruing to rural households living near the river in its entire course of 2525 kilometres, but this study did not measure those benefits. The study was undertaken by the IEG and Metroeconomica.

Chapter 7 looks at the health benefits of improving water quality. Previous studies have come up with some general benefits but they are not attributable to particular kinds of water quality improvement; in particular they do not focus on river water quality. Moreover, they do not establish a quantitative link between water quality and health benefits. Nevertheless, their values are instructive and are reviewed. In this study, the All India Institute of Hygiene and Public Health (AIIH&PH) estimated the benefits of an improvement in water quality as a result of the GAP by comparing the health of communities close to the river, pre- and post-GAP. It also looked at the health of sewerage workers.

Chapter 8 examines the effects of changes in treated/untreated wastewater toxicants (metals and pesticides) discharged by Sewage Treatment Plants (STPs) under the GAP on public health, agriculture, and environmental quality in the disposal (receiving) areas. Although they have some important findings, the impacts in this area could not be quantified in money terms. This work was carried out by the Industrial Toxicology Research Centre.

Chapter 9 reports on the fisheries in the river Ganga. These have been an important resource and continue to be so. Ideally, one would have liked to analyse the fish catch in the river and estimate the change in that catch that could be attributed to improved water quality. In practice this was impossible to do, as the data required for such a sophisticated exercise were simply not available. Not only are the time series very short, they are incomplete for some stations and the corresponding data on effort are also incomplete. Hence, a partial quantitative analysis of changes in the fisheries situation was undertaken without monetary valuation. The main work was carried out by the Central Inland Capture Fisheries Research Institute under the Indian Council of Agricultural Research (ICAR) at Barrackpore.

Chapter 10 reports on the agricultural benefits of GAP which come in three forms. First, there are the irrigation benefits arising from the partially treated water that is released to farms in the GAP area. These were estimated from a survey comparing the cropping patterns, input use, sources of irrigation, and yields for farms receiving the GAP water with similar farms not receiving the GAP water. The second benefit is from the fertilizer value of the irrigation water. It is observed that GAP farms apply less fertilizer than non-GAP farms. Finally, there are the benefits of the use of sludge provided by the STPs. All three benefits are estimated and quantified in money terms. The work was carried out by the IEG.

Chapter 11 examines the impacts of the GAP on the biodiversity of the Ganga ecosystem. It looks at three areas: species of international conservation significance; species that have commercial significance; and overall assessment of 'ecosystem' health. The report provides a judgmental and quantitative review of the impacts of the GAP, but no monetary estimation has been possible. The work was carried out by Dr R.K. Sinha of the University of Patna.

Chapter 12 brings together the cost and benefit data to look at the net benefits, in accordance with the methodology outlined in Chapter 3. It evaluates the net benefits of the two phases of the Ganga programme. The costs of each phase are taken from Chapter 4 and the monetary values are taken from Chapter 5 (non-user benefits), Chapter 6 (user benefits), Chapter 7 (health), and Chapter 10 (agriculture). We acknowledge that the benefits of biodiversity and fisheries could not be quantified in monetary terms. The impacts of toxic changes were less clear but they too have not been quantified. Hence, the analysis is partial to the extent that these have been ignored. The work was carried out by Metroeconomica.

The chapter also reports on how the net benefits change when account is taken of the distributional impacts of the GAP. The cleaning of a very important river like the Ganges in the Indian subcontinent can produce significant income distribution benefits in the Indian economy. However, from the point of view of equity, benefits that accrue to the poor assume greater importance than those to the rich. Therefore, the quantification of income distribution effects is important in the social cost benefit analysis of the GAP. The method employed to estimate the distributional gains and losses was to weigh them according to who bears them. The distribution weights were estimated using a methodology developed by project appraisal economists and applied in many cost–benefit studies. The work was carried out by the IEG and Metroeconomica.

An important issue further dealt within Chapter 12 is, how to finance the GAP on a sustainable basis. A number of different mechanisms were looked at: a polluter-pays principle; a user-pays principle (with government involvement); a user-pays principle (without government involvement); and funding from the general tax system. The analysis was carried out by the IEG and Metroeconomica.

Chapter 13 provides some conclusions and recommendations.

Appendix A provides a formal analysis of the water quality data. The information from the water samples collected at 27 stations is used to calibrate a model which predicts water quality all along the river. This model allows us to estimate the Ganga River quality during 1995, and beyond, both with and without the Ganga Action Plan, Phase-I. The water quality modelling of the Ganga River is used to develop the river water quality scenarios (with particular reference to the Dissolved Oxygen and Biochemical Oxygen Demand) for the year 1995, both with and without the GAP. These are then used to estimate benefits as a function of water quality in Chapters 6 and 7. The work was carried out by the Industrial Toxicology Research Centre at Lucknow working with the National River Conservation Directorate (NRCD).[2]

[2]Comments from the Ganga Directorate and the UK Department for International Development were useful in the development of this work. Their inputs are gratefully acknowleged.

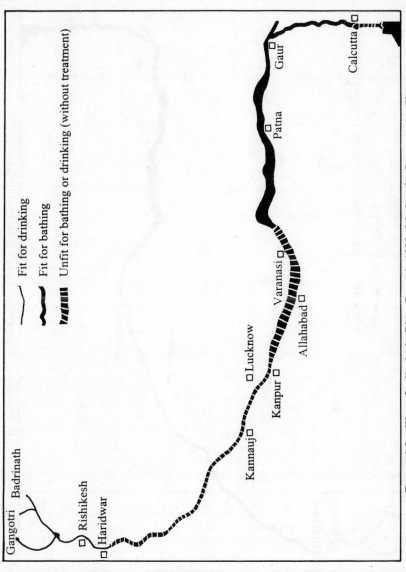

FIGURE 1.2: Water Quality in the River Ganga in 1985–6, before the Ganga Action Plan

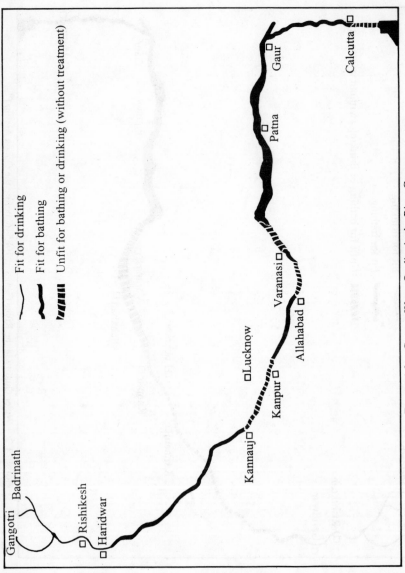

Figure 1.3: Current Water Quality in the River Ganga.

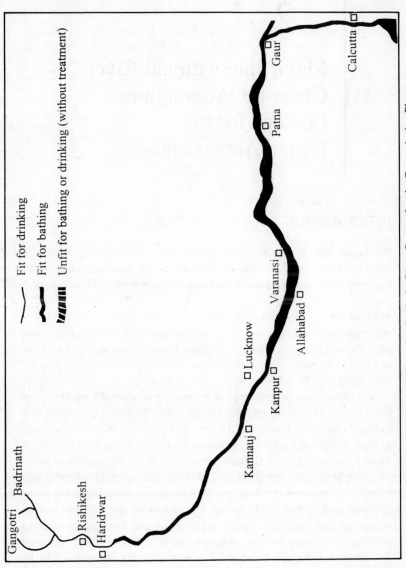

FIGURE 1.4: Expected Water Quality in the River Ganga after the Ganga Action Plan

2

Major International River Clean-up Programmes: Lessons for the Ganga Action Plan

INTRODUCTION

We begin with a review of the major river clean-up programmes in other countries, so as to draw out lessons for the Ganga river. We start by setting the main features of the Ganga river and the actions that were mandated by the late Prime Ministers, Indira Gandhi and Rajiv Gandhi, to clean up the river. The high political profile given to the project is not uncommon for river clean-up programmes; nor is the fact that targets, defined in physical terms, were not based on any detailed analysis of costs and benefits, but were derived from a general feeling that the river needed to be 'cleaned up'.

The three river programmes that have been reviewed are those for the Thames, the Rhine and the Danube. The Thames programme is a national river programme (unlike the other two, which are international in their design and activity). Pollution problems in the Thames have a history that goes back to Roman times. Action had been taken at times of crises in the past, but after 1970 a concerted action programme was launched to improve the river quality. The targets were defined in terms of water quality that would permit fish to survive and to migrate at all stages of the tide. No formal analysis of the benefits of such a programme was carried out, although the 1972 Royal Commission on Environmental Pollution decreed that measures should, in principle, be taken to the point at which the additional costs to the community equal the additional benefits of reduced damages. The river has been cleaned up and now meets the objectives set in 1972, but at a great cost and over a time span of over 20 years.

The Rhine river basin is one of the most industrialized areas in the world and has also been one of the most polluted. The major programmes for its clean-up started in 1950. In 1961 an International Commission of five countries bordering the river—Germany, France, the Netherlands, Switzerland, and Luxembourg—was established to monitor river quality and propose action to clean the river. It had no power to undertake any measures, and so, river quality continued to get worse till the early 1970s. Various actions have been taken since then, with goals defined in terms of lists of substances—a 'black' list of pollutants to be eliminated and a 'grey' list of pollutants to be reduced to meet defined water quality standards. In 1987 the 'Rhine Action Programme' was initiated with targets to be met by the year 2000. These are defined in terms of ecosystem integrity and involve substantial reductions in industrial pollutants. One important goal is to restore salmon to the river by the year 2000. The lessons from this river are the long time it has taken to reach this level and the high cost involved. No discussion, however, has been framed in terms of costs and benefits.

The Danube is also an international river, 2857 kilometres long (longer than the Ganga) and serving 11 countries. It has a great 'self-purification' potential and can cope with the present levels of organic waste. The problems are more to do with nitrate pollution and eutrophication, microbial contamination, and contamination by hazardous substances. Pollution from the river is a cause of serious problems with fisheries in the Black Sea. Concerted action to control pollution in the Danube basin was started in 1985 but progress was limited until recent years, because of the socio-economic and political problems facing many of the countries involved (mainly the former centrally planned economies of Eastern Europe). Much work is now being supported with international funds, starting with the drawing up of a strategic plan. Actions are divided into short, medium, and long term and priorities are being established by looking at the costs and benefits of different sub-programmes. In this respect the Danube river programme differs from the others. 'Hot spots' have been identified and will be acted on urgently. The estimation of the benefits is still in its infancy, but at least work has started on this.

SOME GENERAL PRINCIPLES

There are important lessons to be learnt from the experience of various national and international river clean-up programmes. These lessons have been neatly encapsulated in the principles for river quality management derived from the experience of the engineer, K.R. Imhoff, in the River Ruhr Valley in Germany.[1] The ten principles are:

(i) *Avoidance comes before repair:* Prevention of entry of pollutants into the water system is always preferable to treatment after entry (that is detoxification at the site of production, not at the sewerage treatment works).

(ii) *Kill the big pigs first:* A full balance account of relative share of sources, sinks, and effects of pollutants should precede economic investments, recognizing the fact that there are considerable economies of scale in most waste treatment processes.

(iii) *Retention does not mean elimination:* The removal of contaminants from water through absorption into sediments, as occurs with heavy metals in the River Rhine, does not constitute a long-term solution to river pollution.

(iv) *Dilution is no solution for pollution:* This is true for total pollutant load, but toxic impacts are determined more by concentration than total load and pollution control mechanisms should be sensitive to the potential for severe impacts from high concentration of pollutants developing in low volumes of receiving water.

(v) *Compartment shifting is no final solution:* A focus on water quality alone may simply lead to toxic pollutants being shifted to another component of the overall environment (emissions to air through incineration or to soil by application of contaminated sludge).

(vi) *Redundancy increases safety:* All single point treatment systems can fail, so strategies for treatment at multiple sites can prevent the worst pollution events arising from accidents or failure.

(vii) *Supervision increases safety:* A combination of continuous self-regulation through monitoring, coupled with external regulation to en-

---

[1]See Imhoff *et al* (1991).

sure compliance, is required in order to achieve the highest level of efficiency of waste treatment plant operation.

(viii) *Polluter pays:* The benefits of a clean environment are diffuse and not easily captured by markets. The costs of pollutant release are also diffuse and not paid solely by the polluter. Pollution taxes and other levies are required to ensure that the immediate cost of pollution exceeds the cost of purification.

(ix) *Quality objectives and legal standards are necessary as a challenge for water quality management:* Although pollution standards are required, unrealistic standards, beyond background levels, or beyond what can be measured or achieved with reasonable cost using existing technology can be counterproductive and economically inefficient.

(x) *Multiple use needs partnership:* Enforcement of pollution control standards is extremely expensive; a measure of cooperation and collaboration is required for efficient, sustainable use of river resources, either as sinks or sources.

These principles can provide a general framework against which the success, or failure, of large-scale river basin clean-up programmes can be assessed. Only three such programmes match the scale of the challenge to clean-up the Ganga—the Thames, the Rhine, and the Danube. The clean-up programme for each of these rivers is summarized below before some general lessons are drawn with respect to the Ganga Action Plan.

THE RESTORATION OF THE TIDAL THAMES

INTRODUCTION

The River Thames in England flows as a fresh water river for 245 kilometres from its source near Cirencester in Gloucester, England to Teddington, which is 31.5 kilometres above the present day London Bridge. A weir at Teddington halts the upstream penetration of sea water from the estuary. The tidal Thames runs a total of 151.5 kilometres from Teddington to the outer limit of the old Port of London Authority, 120 kilometres below London Bridge (see Figure 2.1).

The formation of the city of London in Roman times, over the period AD 43–409, was undoubtedly due to the presence of the river. Subsequently, the development of the city of London has been the single most important factor shaping the ecology of the River Thames. Over

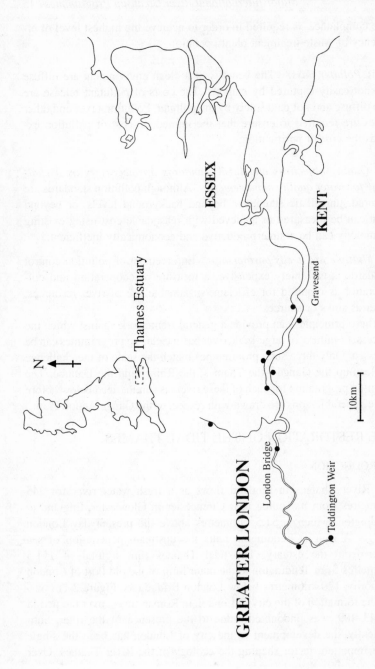

FIGURE 2.1: Map of the Tidal Thames River

the next 1500 year, the natural flow and ecology of the river basin has been transformed in many ways, the most significant of which are the reductions in the inflow from tributaries in the London area and the dramatic changes in the load of inputs arising from human activities within the basin.

These changes have led to two major pollution crises in recent history (that is the last 150 years). In both of these crises, long stretches of the tidal Thames became anaerobic and 'ecologically dead', yet in both cases it has been possible to restore the river to a state resembling its previous condition.

The first 'crisis' arising over the quality of water in the Thames arose not from a concern about ecology, but rather as the result of its impact on human health through an increase in cholera outbreaks. This was caused by abstraction of drinking water directly from the river, which was receiving greatly increased amounts of untreated sewage following the introduction of legislation, in 1815, requiring cesspools to be connected to the sewers.

During the period 1815–50 the quantity of raw sewage entering the river continued to grow in volume, especially as the result of the growing popularity of water closets, the installation of which was made mandatory for all new houses after 1850. This polluted state of the river reached its peak in 1858, the 'Year of the Great Stink', when disinfectant-soaked sheets were hung in the windows of the Houses of Parliament to reduce the smell.

This 'direct action' by the river succeeded in breaking an impasse which had stymied the attempts of successive Royal Commissions to address the increasing pollution of the river by sewage inflow. Options to utilize sewage solids as an agricultural resource were rejected in favour of diversion of waste in pipes to holding points downstream (at Beckton and Crossness), where release into the ebbing tide would remove it from the immediate vicinity of London.

These measures did result in a temporary improvement in the condition of the river at London, but the improvement was soon negated by the continued increase in total discharge into the river and the emergence of additional pollution problems downstream. Thus, in spite of the major improvements made in collection, treatment, and disposal of London's sewage, the ecological condition of the tidal Thames declined rapidly from 1910 to 1950, by which time the river was almost completely anaerobic for over 40 kilometres of its length within London.

This decline in river quality was the result of a set of changes, all happening simultaneously, over the period 1935–50. Of these, the most significant were:

- increase in effluent output from the two major outfalls at Beckton and Crossness;
- additional effluent output from a number of smaller additional sewage works constructed to cope with a growing and geographically spreading population;
- inputs from a wide range of non-point sources (for example agricultural land);
- industrial discharges and thermal pollution from power stations;
- storm flows;
- greatly increased discharge of non-biodegradable synthetic detergents.

The overall consequence of all these changes was to overload the ability of the river to receive pollution (that is to load it to the point where it could not self-purify and where dissolved oxygen reserves were reduced below 5 per cent of saturation levels). At this low point all normal biological life is eliminated and the only remaining significant ecological processes are anaerobic bacterial chemosynthetic reactions which oxidize nitrates and sulphates to nitrogen and highly toxic (hydrogen) sulphides, respectively.

This collapse of the ecology of the river coincided with the Festival of Britain, when many foreign boats were moored along the river. Hydrogen sulphide fumes released from the anaerobic muds of the river blackened and tarnished the metal and paint work of the ships in spite of attempts to sterilize the river by the addition of chlorine to the effluent at Beckton. Once again, direct impact on humans, especially those holding positions of power, proved crucial in bringing about concerted action to improve the condition of the river.

THE SECOND RESTORATION OF THE RIVER THAMES 1950–80

The success of the second major rehabilitation of the tidal Thames owes much to the enthusiasm of the responsible bodies and the quality of scientific information that was available as the result of long-term regular monitoring initiated in 1893. This provided the foundation upon which the Water Pollution Research Laboratory could build its study, finally released as *The Effects of Polluting Discharges on the Thames Estuary* in 1964.

The beginning of the second restoration can be traced to the activities following from the recommendations of the Pippard Committee report of 1961. However, even before the recommendations of this committee had been fully implemented, the Third Report of the Royal Commission on Environmental Pollution, published in 1972, broadened and extended the scope of environmental recovery, both for the Thames and other rivers in the UK.

The Commission adopted the pragmatic position that there was a practical limit to the burden that could be imposed upon the community for the abatement of pollution of estuaries. This limit should be the point at which the marginal cost of abatement equalled the marginal cost of the damage done (the classical Cost Benefit Analysis optimum, as discussed in Chapter 3). But, because such damage costs were not readily available, the Commission made a more pragmatic series of recommendations, namely,

- pollutants not rendered harmless by natural processes, and which accumulate in benthic mud or in living organisms, should be removed from effluents before they are discharged;
- the quality of the estuary should be maintained by the use of pollution budgets, calculated from monitoring data, which would allow calculation of the maximum pollutant load compatible with a particular quality target.

Initially, these targets were outlined as follows:
- to exploit the estuary for waste disposal up to a level which does not endanger aquatic life, or transgress the standards of amenity which the public need and are prepared to pay for;
- to ensure, by controlling quality standards, that exploitation of the estuary does not exceed this level; and
- the policy of planned pollution budgets would require close co-operation between planning authorities and river authorities (or their successors), and there should be consultation on any relevant plans.

With respect to the requirement to 'not endanger aquatic life', the Commission suggested two simple biological criteria which could be used for monitoring purposes:

- to support on the mud bottom the fauna essential for supporting sea fisheries;

- to allow the passage of migratory fish at all stages of the tide. The quality needed for migratory fish was considered to be a dissolved oxygen content of not less than 30 per cent saturation in April/May when smelts pass downstream to the sea (in 9 out of 10 years) and a temperature of less than 20°C.

Thus, for the first time, the Commission recognized the benefit of introducing environmental quality objectives (EQOs) rather than specific fixed emission standards.

With regard to authority, the commission recommended that pollution control of a particular estuary should lie with a single authority; in the case of the Thames this would be the regional water authority (Thames Water Authority).

Although in retrospect these measures seem rather obvious, it was the first time that fundamental shifts in approach had been adopted on such a large scale and by such an eminent authority. These shifts in policy, especially the setting of location-specific environmental quality objectives, based on detailed scientific monitoring (and modelling) under the jurisdiction of one major authority, have become the foundation for all modern aquatic pollution control.

ENVIRONMENTAL QUALITY OBJECTIVES (EQOs)

The adoption of specific EQOs for different sections of the tidal Thames in the 1980s was an important progressive feature of the Second Restoration Plan for the Tidal Thames (see Table 2.1). Ironically, the EQO adopted, that of making the river suitable for the passage of migratory fish, was exactly that which was proposed by the Thames Survey committee and rejected as being beyond reasonable economic worth by the Pippard Committee in 1961.

Achieving these EQOs requires the adoption of a pollution budget which would restrict the pollution load applied to any stretch of the river to an amount which would not reduce dissolved oxygen levels below those set out in the EQO. This amount could be predicted by the mathematical model developed by the Water Pollution Research Laboratory and although it could, in principle, be achieved by many different combinations of effluent flow, practical considerations of existing capital investments reduced the number of options for action to the following activities:

TABLE 2.1
River Water Quality Objectives for Sections of the Thames

River Section	*Quality objective*
Teddington to London Bridge (31.5 km)	(i) the minimum percentage of air saturation with dissolved oxygen should be at least 40 per cent in the case of 95 per cent of the samples analysed;
	(ii) the water should be non-toxic to fish.
London Bridge to Canway Island (65 km)	(i) the minimum percentage air saturation with dissolved oxygen should be at least 10 per cent in the case of 95 per cent of the samples analysed;
	(ii) the quarterly minimum average percentage saturation with dissolved oxygen should not fall below 30 per cent.
Canway Island to seaward limits (55 km)	(i) the quality should be suitable for the whole life cycle of marine organisms, including fish;
	(ii) the minimum percentage air saturation with dissolved oxygen should be at least 60 per cent in the case of 95 per cent of the samples analysed.

- improvement[2] of Beckton works[3] to contain the emission standard of Biological Oxygen Demand (BOD) at 10 milligrams per litre and of ammonia at 2.5 milligrams per litre. This would yield consider-

[2] Dissolved oxygen saturation of 35 per cent in May corresponds to a minimum average of 30 per cent in the third quarter of the year (July–September) when oxygen levels are at their worst. This, in turn, corresponds to a 95 percentile value of no less than 10 per cent dissolved oxygen saturation in all parts of the river; this was therefore adopted as a standard for the most polluted sections of the river with higher standards being set for other sections of the river.

[3] The Beckton works were originally constructed in 1932 and have been progressively expanded and upgraded over the past 40 years, including a £7.5 million extension in 1959, and a further additional treatment plant capable of treating 1.14 million m^3/day at a cost of £21 million in 1967. The final upgrading added eight new primary sedimentation tanks (each of 11,560 m^3 capacity), eight diffused, air-activated sludge tanks of the 'single pass' type (each of 27,000 m^3 capacity), and 48 further final sedimentation tanks (each of 3500 m^3 capacity). In addition, the existing sludge digestion plant was extended by providing 32 additional tanks (each of 4750 m^3 capacity); these improvements were completed in 1974.

able improvements in river quality, even with unchanged conditions at other works. Beckton was, in 1982, still the largest sewage treatment works in Europe, capable of receiving a flow of 2.73 million cubic metres a day and producing a volume of effluent equal to that of the River Medway, the largest tributary of the Thames.

- improvement of Long Reach (West Kent) works from an existing emission standard of BOD at 150 milligrams per litre and ammonia at 32 milligrams per litre to BOD at 20 milligrams per litre and ammonia at 30 milligrams per litre (non-nitrifying); improvements completed using diffused-air activated sludge treatment by the end of 1979.

- restriction on thermal discharge from any existing or planned industrial plant upstream of Beckton works; any power station which needed to be located on the river above this point would need a cooling tower to ensure that its discharges of cooling water did not raise the river temperature above that compatible with the EQO of a temperature of less than 20°C.

- the Riverside works has a total flow of only 94,000 cubic metres a day but received effluent from a number of factories including one which was discharging compounds which would inhibit the oxidation of ammonia in any activated sludge process.

CONCLUSIONS:

The successful clean-up of the Thames shows that it is possible to redeem even an extremely polluted aquatic ecosystem. The Thames is of paramount importance, providing a sink for approximately one-third of the sewage of Great Britain. In dry weather the flow of sewage effluents entering the estuary is four times as great as the freshwater flow over Teddington weir, just above London. The major aim of the Royal Commission on Environmental Pollution (1972), which was to exploit the estuary for waste disposal up to a level which does not endanger aquatic life, has been achieved. Not only has this involved an enormous amount of foresight and scientific investigation, but also the expenditure of large sums of money.

At least £100 million has been spent over the last 20 years or so to transform the estuary from an anaerobic tidal channel, through Greater London and further downstream, to an aerobic tideway supporting normal fisheries and able to support the passage of migratory salmon, as well as to provide many other indirect environmental benefits which are difficult to quantify. For example, during the drought of 1976, virtually all freshwater flow of the Thames was taken for public supply, giving an additional yield of some 179 million gallons a day. If the estuary had not been cleaned up, none of this additional yield could have been taken without enveloping riverside London in a noxious and poisonous atmosphere of hydrogen sulphide of unprecedented and unacceptable proportions. In the event, the estuary, with no significant freshwater input over many months, reached its highest degree of cleanliness in almost a century. This demonstrates that the estuary clean-up had given, and can continue to give when necessary, an extra massive yield of freshwater resources for public supply, and thus shows how the improvement in estuary quality can more than pay for itself as a water conservation measure.

Although the restoration of the tidal Thames has become the model for successful river clean-up programmes, not all of the lessons learnt can be easily applied to other systems. In practice the Thames situation was relatively simple to measure, model, and predict, because of the small geographical scale, the relatively small number of major pollutant inputs, and the high level of technical expertise and capital which could be applied to the problem, once the political will was established. The problem with the Ganga is spread over a huge geographical area, with a cumulative sequence of pollutant inputs arising from a much greater diversity of sources, and is subject to much more extreme variations in flow characteristics, both within and between years. Although many of the lessons learnt from the Thames were applied to the Ganga problem through the direct involvement of Thames Water International, it is not surprising that a 5 year plan did not achieve as much for the Ganga as 25 years of continuous investment in water quality improvement has for the Thames.

THE DECLINE AND ECOLOGICAL RESTORATION OF THE RIVER RHINE

GEOGRAPHY

The River Rhine runs a total of 1320 kilometres from its headwaters in the Swiss Alps, through Lake Constance, the Black Forest, and on through Germany, France, and the Netherlands before discharging into the North Sea (See Figure 2.2). The catchment area of the river covers

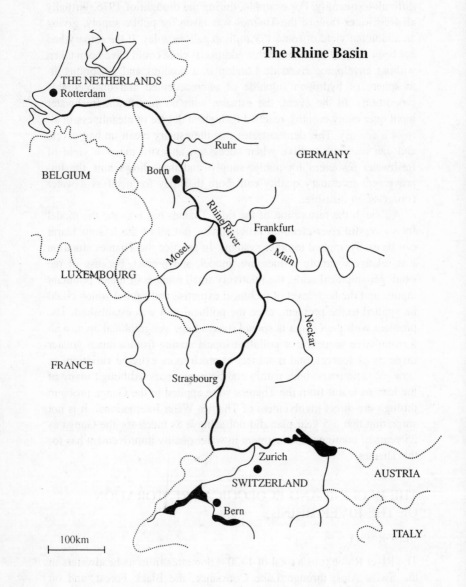

FIGURE 2.2: Map of the Rhine River and Basin

almost 200,000 square kilometres, and has a population of over 50 million people, including most of Switzerland, the south-western provinces of Germany, the north-eastern corner of France, all of Luxembourg, and most of the Netherlands. Around 20 million people derive their drinking water from the river (including Lake Constance) and around 20 per cent of the total flow is diverted for consumption or industrial use. Although not the largest European river, the Rhine basin is the focus of perhaps the highest density of industrial activity that has ever existed, accounting for around 10 to 20 per cent of the total chemical industry production in OECD countries, and 10 hydroelectric plants generating a total of 8.7 megawatt-hours per year. In addition, the Rhine basin is a centre of agricultural production, with one of the highest yields but also the highest levels of agrochemical inputs in the world.

RECENT HISTORY OF WATER QUALITY CONTROL

The enormous changes which have taken place along the river Rhine and the extent to which these changes have altered the ecology of the river have required a major, and enormously expensive, effort in order to halt and reverse the decline in water quality which had taken place in the twentieth century. The precise role played by any one set of political actors, legislation, or clean-up activities cannot be isolated completely from the many others which preceded, overlapped, or followed them. Indeed, some of the most significant water quality improvements have arisen from actions taken for other reasons, independent of the condition of the Rhine (for example those following legislation and technological changes designed to reduce industrial emissions to air).

Of the many institutional measures designed specifically to improve the water quality of the Rhine, the most important have been the three main international programmes developed since 1950. These are described in more detail below.

THE INTERNATIONAL COMMISSION FOR THE PROTECTION OF THE RHINE AGAINST POLLUTION (ICPR)

In 1950, the countries bordering the Rhine joined together to discuss the serious decline in the Rhine river water quality. However, the formal establishment of a commission was not completed until the International Commission for the Protection of the Rhine against Pollution (ICPR) was created by an agreement signed at Berne, on 29 April 1963, by five riparian states: the Federal Republic of Germany, France, the Netherlands, Switzerland, and Luxembourg. The 1963 Berne Convention was a purely institutional arrangement and contained no specific obligation for the contracting parties beyond their agreement to cooperate with the Commission to work towards the following objectives:

- to have a detailed analysis of pollution of the Rhine (nature, extent and source) and evaluate the results;
- propose actions to protect the Rhine;
- prepare international treaties.

Because the only task of the Commission was to monitor the pollution of the Rhine, without having the power to make any decisions, pollution of the river continued to worsen during the 1960s and early 1970s.

By the early 1970s the self-purifying capacity of the river had been exceeded by the large quantities of untreated organic waste water entering the river. Long stretches of the river—for example the 200 kilometres stretch from the mouth of the River Main to Cologne—were almost completely deoxygenated. Almost every stretch of the river had suffered a loss of species and a shift in species composition to those tolerant of polluted conditions, and the number of heavy-metal contaminated and diseased fish was at an all-time high.

In the summer of 1971, the middle course of the Rhine was entirely without oxygen along a stretch of over 100 kilometres, a situation that ultimately prompted European governments to take action. On 22 March 22, 1972, the European Community requested that the parties to the Berne Convention design an emergency programme for decontaminating the Rhine. On 20 June 1975, the European Parliament passed a resolution, calling on those countries that were simultaneously, parties to the Berne Convention and members of the European Community, to agree on 'immediate, practical and co-ordinated measures to avoid the impending disaster'. Finally, at a conference called by the Netherlands between the ministers of parties to the Berne Convention, the ICPR presented a long-term programme of activities. The programme contained a description of the hydrological character and current quality of the waters of the Rhine, as well as data concerning current and foreseeable sources of pollution. The last part of the programme was a plan of action for the Commission and outlined future activities in some principal fields, including pollution by sewage waters, radioactive substances, chemicals, chlorides, ships, thermal pollution, and emergency warning systems. Progress within each of the principal fields has varied considerably, with separate agreements and activities for some fields, achieving different levels of success. In some fields, for example thermal pollution, little real progress has been made within the ICPR.

THE CONVENTION ON CHEMICAL POLLUTION (1976)

The ICPR programme was the framework around which the Convention on the Protection of the Rhine against Chemical Pollution was based. This convention was adopted by the member states on 25 May 1976. In fact, the main substantive output from the 1976 meeting was the Convention on The Protection of the Rhine Against Chemical Pollution, signed by the ICPR contracting parties. This was an outline convention which provided for the setting-up of threshold values for the discharge of individual noxious substances into the environment of the Rhine basin. It corresponded for the most part with the EC Directive 76/464/EEC of 4 May 1976 on 'Pollution Caused by Certain Dangerous Substances Discharged into the Aquatic Environment of the Community', which identified 132 substances requiring regulation. Recommendations passed within the convention must become part of national law and become legally binding once all contracting parties have ratified them.

The Convention on the Protection of the Rhine Against Chemical Pollution amended and modified the Berne Convention by placing additional responsibilities on the International Commission and the Contracting Parties. Article I of the Convention sets two goals for improving the water quality of the Rhine:

- elimination of pollution of the Rhine by certain highly dangerous substances, enumerated in the 'black list' of Annex I of the Convention and
- reduction of pollution of the Rhine by substances listed in the 'grey list' of Annex II of the Convention.

The limits for discharges of Annex I substances are set by an international body, rather than a national one, albeit by the unanimous agreement of the contracting parties. As part of the 'first step' in this process, the governments of the contracting parties to the Convention agreed to compile an inventory of discharge of substances mentioned in Annex I and to communicate the contents thereof to the International Commission. The inventories must be updated at least every three years. Prior authorization must be obtained from the component government entity for any discharges of Annex I substances into the surface waters of the Rhine basin. This authorization fixes the emission standards, including norms and time limits governing such discharges that cannot exceed the limit values agreed upon by the parties to the Convention.

The Convention established a list of 83 priority substances (Annex I of the Convention) which it identified as being significant threats to water quality on the basis of environmental persistence, toxicity, potential for bioaccumulation in living organisms, and the total amounts produced or processed.

The competent national authorities should only authorize those discharges which at least respect the threshold values for concentration of individual substances fixed in Appendix IV of the Convention on chemical pollution. These threshold values are applicable to the waste water discharged into a surface water body by a complete industrial site. The threshold values are based on the available techniques of waste water treatment (basically, 'end of pipe' controls), but also take into account modifications of production techniques and on-site measures applicable to waste water.

Between 1980 and 1987, the ICPR has worked on 20 of the 83 listed substances; for 12 of these threshold values for discharge were fixed, whilst the other 8 were 'placed under surveillance' until additional discharge data became available. On average, it has taken 6 years for the proposed threshold values to be ratified and even longer for abatement measures to be applied. During this time, other national or international limits on release of pollutants have contributed to the reduction in pollution load on the Rhine, and have limited the effective use of the powers of the convention.

Throughout this period, the water quality monitoring programme on the River Rhine has continued to develop, and a large quantity of data are collected and published each year by the ICPR (in German and French only). By 1992, the last year for which data have been published, there were 11 monitoring stations along the river, collecting data on up to 58 chemical or physical parameters, including 23 organic micropollutants (atrizine, endosulphan, benzene, etc.), 10 heavy metals (nickel, cadmium, arsenic, etc.), 6 inorganic mineral ions (chloride, sulphate, calcium, etc.), 6 measures of eutrophication (phosphate, nitrate, ammonia, etc.), 4 measures of radioactivity (potassium-40, tritium, etc.), and 6 physical measures (flow, temperature, pH, etc.). By 1992, a total of around 40 of these substances had water quality standards (objectives). These standards are listed in Table 2.2.

TABLE 2.2
Recommended Water Quality Standards for the River Rhine

No.	Pollutant substance	Standard	Authority
1.	Chloronitrobenzenes	1 µg/1	IRC
2.	Trichlobenzene	0.1 µg/1	IRC
3.	Pentachlorophenol	0.1 µg/1	IRC
4.	Trichloroethene (trichloroethylene)	1 µg/1	IRC
5.	Tetrachloroethane (perchloroethylene)	1 µg/1	IRC
6.	Chloroanilines	0.05 µg/1	IRC
7.	Parathion	0.0002 µg/1	IRC
8.	Benzene	2 µg/1	IRC
9.	1,1,1-trichloroethane	1 µg/1	IRC
10.	1,2-dichloroethane	1 µg/l	IRC
11.	Azynophos-methyl	0.001 µg/l	IRC
12.	Bentazon	0.1 µg/l	IRC
13.	Simazin	0.06 µg/l	IRC
14.	Atrazin	0.1 µg/l	IRC
15.	Dichlorovos	0.7 µg/l	IRC
16.	2-chlorotoluene	1 µg/l	IRC
17.	4-chlorotoluene	1 µg/l	IRC
18.	Organotin compounds	0.001 µg/l	IRC
19.	Trifluralin	0.002 µg/l	IRC
20.	Fenthion	0.007 µg/l	IRC
21.	Mercury	0.02 µg/l	IRC
22.	Cadmium	0.03 µg/l	IRC*

No.	Pollutant substance	Standard	Authority
23.	Chromium	3.21 µg/l	IRC*
24.	Copper	2.92 µg/l	IRC*
25.	Nickel	2.08 µg/l	IRC*
26.	Zinc	7.86 µg/l	IRC*
27.	Lead	2.7 µg/l	IRC*
28.	Carbon tetrachloride	1 µg/l	IRC
29.	Chloroform	0.6 µg/l	IRC
30.	PCB	0.0001 µg/l	IRC
31.	Aldrin, dieldrin, endrin, isodrin	0.001 µg/l	IRC
32.	Endosulphan	0.001 µg/l	IRC
33.	Hexachlorobenzene	0.001 µg/l	IRC
34.	Hexachlorobutadiene	0.5 µg/l	IRC
35.	Phosphates	0.15 mg/l	IRC
36.	Ammonium	0.2 mg/l	IRC
37.	Nitrate (+ Nitrite)	10 mg/l	IMP
38.	Kjeldahl Nitrogen	None	
39.	t-N (calculated)	2.2 mg/l	AMK
40.	Total Nitrogen	2.2 mg/l	AMK
41.	AOX/EOCl	50 µg/l	IRC
42.	PCAH's	0.0003 µg/l	AMK

*Based on standard for suspended solids.

Note: µg/l: micrograms in a litre

mg/l: milligrams in a litre.

It should be stressed that, although there were major improvements in some important aspects of water quality during the period 1970–85 (for example see Table 2.3 for Bimmen), the water quality standards listed in Table 2.2 are still not achieved for much of the time at a majority of the water quality monitoring stations. Furthermore, they are unlikely to be met in the near future according to preliminary modelling (see the following pages).

TABLE 2.3
Improvement in Water Quality of the River Rhine
at Bimmen, Germany

Parameter	1978	1988
BOD5 (mg/l)	4.5	3.0
COD (mg/l)	22.0	18.0
Ammonium (mg/l)	0.9	0.3
Nitrate (mg/l)	3.7	3.8
Phosphorus (mg/l)	0.6	0.3
Cadmium (μg/l)	2.0	0.25

BOD5: Biological Oxygen Demand at 5 days.
COD: Chemical Oxygen Demand.

It is difficult to estimate the total amount spent to achieve the improvements in river water quality and no disaggregated economic data
are available. Between 1965 and 1989, it is estimated that the member
states of the International Commission for the Protection of the Rhine
against Pollution (ICPR) spent about DM 100 thousand million on improving water quality, primarily through the construction of new waste
water treatment plants and the enlargement of existing ones. This expenditure has produced some noticeable improvements, especially with
regard to dissolved oxygen levels. The mechanism by which this money
has been raised has varied between member states, but in all cases, some
part of the cost has been borne by a wastewater levy of some kind. The
ICPR has compared the different systems of wastewater levy existing in
the states of the contracting parties. It has concluded that they are efficient, with whoever discharges wastewater into the sewage system
having to pay for it in all five member states, so that the public authorities
may maintain and expand the sewage system as well as the wastewater
treatment plants. If industrial plants or municipalities discharge directly
into brooks and rivers, a sewage tax is levied by the Netherlands, France,
and Germany. This is not yet the case in Switzerland and Luxembourg
where waste water levy systems are under discussion (as of 1994).

Between 1975 and 1985 the pollution of the Rhine by oxygen deplet-
ing substances was reduced by almost 60 per cent. In the same period,

oxygen saturation increased considerably, from about 60 to 84 per cent, and has reached as much as 93 per cent in the 1990s. Since 1990, the oxygen content has not sunk below the mean value of 9.6 milligram per litre (minimum 5 milligram per litre; maximum 13 milligram per litre), consistently above the minimum of at least 4 milligram per litre which fish require for normal activity. Further details of quality changes are given in Table 2.3.

THE RHINE ACTION PROGRAMME (RAP)

Interestingly, it was a major industrial accident that generated a further reawakening of public interest in renewed activity to improve the water quality of the Rhine. On 1 November 1986, a fire at the Sandoz chemical factory at Schweizerhalle, near Basel, resulted in the release of extremely toxic pesticides and mercury compounds into the river, killing thousands of fish and freshwater invertebrates.

Although the biota of the affected stretch of the river recovered substantially within a year of the accident, the public outcry following this accident, with 10,000 people marching through the streets of Basel, protesting about damage to the river Rhine, led to renewed international discussion. The eventual outcome of this renewed activity was the Rhine Action Programme (RAP), which was approved by the 8th Conference of Ministers on the Protection of the Rhine against Pollution, held at Strasbourg on 1 October 1987.

The RAP has the following well defined targets, to be achieved by the year 2000:

- The ecosystem of the Rhine is to be restored as an entity and the water quality is to be restored to such an extent that formerly existing species, such as salmon and sea trout, may return;
- Rhine water is to be continued to be used for drinking water supply;
- the pollution of the river sediments is to be reduced to such an extent that this sludge may at any time be used for landfilling or be dumped in the sea.

In addition, after severe ecological disturbances in the North Sea during 1988, a series of additional goals were adopted to support the goal of the North Sea Conference to stabilize the ecological state of the North Sea.

The specific goals of the RAP were that:

- the discharge of the most important noxious substances into the Rhine is to be cut down by 50 per cent as compared to the 1985 levels;
- levels of safety with regard to waste and accidents in industrial plants are to be tightened;
- a minimum level of surveillance of dischargers is to be installed;
- adequate environmental conditions must be restored for the flora and fauna typical of the Rhine, for salmon and other migrants. Barrage weirs must be equipped with fish passages and spawning grounds must be restored in the upper reaches of the tributaries.

The RAP marked a significant shift in the approach taken to water quality improvements on the Rhine, away from water quality standards for individual specified chemicals to one based upon the best available technology (BAT) which might be adopted in particular branches of industry. This shift was in part due to the success of the previous activities, which had produced a general increase in the ecological health of the river through reduced BOD and COD. In addition, it was prompted by declining concentrations of most halogenated organic compounds in the river, which were approaching the limits of detection. The most important of these industrial branches are:

- pulp paper production*
- surface treatment of metals (galvanization)*
- organic chemistry*
- production of paper and cardboard*
- refineries‡
- production of fertilizers†
- production and refinement of textiles†.

(As of 1994: * BAT already implemented ;   † BAT introduction under discussion;
      ‡ already regulated by other organizations).

The actual BAT to be utilized is not fixed in the treaties, but rather, is a recommendation based on mutual confidence between industry and the ICPR. For each of these industrial sectors listed above, BAT aims to avoid environmental pollution, not just of wastewater, but rather, attempts to:

- avoid displacing pollution from wastewater to other media (solid waste/soils/air);
- avoid technologies producing large volumes of wastewater and chemical procedures polluting the environment;
- encourage full disclosure of amount and composition of all wastewater flows;
- use sum parameters reflecting amounts of total pollutants released rather than concentrations. The use of sum parameters has the advantage of combining more complete coverage of total pollutant load (including by-products and similar non-priority substances) with reduced monitoring effort.

In order to speed up the reduction of 'permanent' pollutants from direct and diffuse sources, the ICPR has adopted the following measures:

- a baseline discharge inventory of priority substances for 1985 (partly 1990);
- a prognosis indicating the expected discharge of priority substances in 1995 (after implementing best available technology (BAT) for point sources and best environmental practice (BEP) for diffuse sources);
- adoption of preliminary quality objectives for priority pollutants.

An example of the new approach is that adopted for AOX (Adsorbable Organic Halides). In 1985, 50 per cent of total AOX in the Rhine came from the paper/pulp industry, mostly above Koblenz. Once released into the river, degradation of these substances is minimal. Between 1985 and 1992, following the introduction of BAT, total AOX discharges were reduced by 80 per cent from about 6700 tonnes to 1200 tonnes. An important part of the BAT reduction has been the elimination of the use of elementary chlorine for bleaching, thus avoiding the production of chloroform waste. Chloroform discharges were reduced from 110 tonnes in 1985 to 3 tonnes in 1992.

An ICPR inventory made in 1992 showed that the point source discharges (of industrial and communal origin) of most priority discharges have been reduced by more than 50 per cent, in many cases by more than 80 per cent, between 1985 and 1992.

The goals set by the ICPR are extremely ambitious and will not be met for all priority substances by the use of current BAT and BEP within the time frame of the project.

Preliminary analyses[4] (A.P. Benoist and G.H. Broseliske, 1994) indicate that the International River Commission (IRC) water quality objectives will be met by 1995 for 15 of the 40 or so priority substances, whilst insufficient data exist for reliable estimates to be made for a further 6 substances.

Even with the application of BAT and BET, water quality objectives will not be met across the entire Rhine basin for approximately 20 (50 per cent) of the priority substances. A crude estimate of the costs of reducing emissions of these 15 substances to the water quality objectives, using technology beyond BAT and BEP in the Netherlands alone, totalled Dutch guilders (Dfl) 100 billion.

Of these 20 substances, it is predicted that the objectives could be met across the entire basin for only three categories of substances: Bentazon, Absorbable Organic Halides (AOX)/Extractable Organic Chlorine (EOCl), and Hexachlorobenzene, at an estimated additional annual cost of Dfl 0.77 million, Dfl 202.0 million, and Dfl 14.3 million respectively, if additional abatements beyond BAT and BEP were to be implemented. This would include the use of sand filters as effluent polishing devices, plus using sand and activated carbon filters at industrial point sources.

It appears that the quality objectives for heavy metals will be almost impossible to achieve because a large part of the quality objective concentration arises from natural sources.

THE ECOLOGICAL MASTER PLAN FOR THE RHINE

In addition to its focus on specific priority pollutants, the RAP has set itself more 'holistic' ecological goals to keep the Rhine ecosystem alive and in good health, and to restore species that have disappeared. The return of the salmon by the year 2000 is a further specific, high profile, and easily measured objective of this programme. Details of how this will be achieved are laid down in the Ecological Master Plan for the Rhine and Salmon 2000 programme, summarized below.

[4]A.P. Benoist and G.H. Broseliske (1994), 'Water Quality Prognosis and Cost Analysis of Pollution Abatement Measures in the Rhine Basin (The River Rhine Project: EVER)' *Water Science Technology,* 29(3), pp. 95–106.

In order to achieve these ecological objectives, the water quality of the Rhine along its entire length must be improved beyond those already achieved over the past 20 years.

In order to achieve the long-term goals of the RAP, the ICPR presented the Ecological Master Plan for the Rhine, which gives a more detailed description of the conditions for the return of the migrants to the Rhine and its tributaries. Perhaps the most significant of these will be that all remaining natural alluvial habitat along the main river is to be protected from any further exploitation, so that it can be used as a 'reservoir' of biodiversity from which recolonization of restored stretches of the river can occur naturally.

The implementation of the RAP is taking place in three distinct phases. Phase I (1988–89) was a blueprint phase in order to plant the abatement measures for Phase 2, which was implemented between 1989 and 1995. Phase 3 is intended to provide additional measures to supplement those of Phase 1 and 2 in the event that these do not achieve the intended objectives.

The 'Salmon 2000' component of the Ecological Master Plan for the Rhine has two main objectives:

- The first objective is that migratory fish (salmon, sea trout, allice shad, sea lamprey, sturgeon) should return to the Rhine. Migratory fish are defined as those species moving from fresh water to salt water and vice versa, and which spawn in the upper reaches of a river system. They require the habitat of the entire river for their life cycle. The salmon is the best-known example and is therefore used as a symbol and flagship species for the entire programme. If conditions are such that the salmon thrives, many other endangered species also profit from the situation. The main stream must again become an efficient habitat for migratory fish, which means that unhindered fish migration upstream towards the spawning grounds and downstream towards the sea must be possible. Since the Rhine has become a waterway for large ships, with obstacles such as barrage weirs, power stations, and dams, free migration is no longer possible. Consequently, if the main stream is to be restored as the backbone of the 'Rhine' ecosystem, the present obstacles must be equipped with properly operating devices, allowing the fish to surmount them (for example fish-ladders and sluices). According to first estimates, costs for the necessary structural modifications will amount to DM 110 million.

- However, not only the river itself as the backbone of the ecosystem, but all connected habitats, such as the riverbed, the banks, and alluvial areas must be restored, thus allowing self-regulating food chains to develop. The loss of alluvial areas is particularly threatening to the ecosystem in the regions of the Upper and Lower Rhine. Due to the construction of dykes and the straightening of the river's course, 90 per cent of the alluvial areas between Basel and Karlsruhe, for example, have vanished, reducing biological diversity within the river system. The remaining alluvial areas and other zones of ecological importance must be preserved, protected, and, wherever possible, extended, in order to restore living conditions for a greater variety of species. The ICPR has completed extensive field surveys which have provided specific information on ecologically important areas along the Rhine which have been, or are to be, placed under protection.

THE CAUSES OF THE LOSS OF MIGRATORY SPECIES

There are three main causes for the reduction or disappearance of the stock of migratory fish in the Rhine. Each of these problems is being addressed within the Ecological Master Plan for the Rhine:

River development and loss of alluvial habitat

Problem: During the last century man began transforming the river into a navigable waterway. Even though fish could still pass upstream and downstream, migration possibilities deteriorated on account of the uniform nature of the channel (lacking diversity) and the absence of resting places. Between 1955 and 1977, more than half the former flooding zones along the Upper Rhine were lost. These zones are now well protected against inundation, and are inhabited and used for farming. Spawning grounds in the river were destroyed and the rapid flood drainage affected the development of spawn. The construction of dykes and bank stabilization systems cut off many meanders, fast-flowing tributaries, and river-meadows. This has resulted in a tremendous loss of spawning grounds (gravel banks) and nursery grounds.

Action: List and restore intact spawning and nursery grounds for salmon and other migratory species. Investigations so far carried out have shown that 150 hectares of habitats fit for restoring the salmon stock still exist in the old bed of the Rhine (Restrhein), 30 hectares in the Breusch and 50 hectares in the Sauer. Other reaches and tributaries are being examined with a view to finding suitable or retrievable habitats. In addition to ensuring that there is no more loss of alluvial habitat to any form of development, there is a possibility that some lost habitat could be regained and restored if the EC programme for land set aside is applied to areas near the Rhine.

Cost: Expenditures of the salmon habitat projects for which subventions were asked: 4.9 million ECU = 10 million DM; (EC subvention: 50 per cent; States (Lander) share: 50 per cent)

Barrage weirs and power plants

Problem: The 11 power plants along the High Rhine (Hochrhein) and a further 10 along the Upper Rhine have disrupted upstream fish migration and barred access to the spawning grounds. Additionally, a significant number of young fish have been killed by the turbines on their migration downstream. With the construction of 14 barrage weirs (1958–64), the Moselle too became a navigable waterway, and fish migration was impeded or hampered on this major tributary of the Rhine.

Action: Barrage weirs, which today disrupt migration, must be made passable. The map gives a general survey of the large number of barrage weirs in the Rhine catchment area which are considered as possible obstacles to migration. The control of the sluices on the Haringvliet and Ijsselmeer (estuary of the Rhine in the Netherlands) is to be adapted to the requirements of migratory fish in order to make it easier for them to surmount these obstacles. The possibilities of surmounting three more barrage weirs along the Nederrijn/Lek are currently under examination. Beyond these barriers, the main channel of the Rhine is unregulated until Karlsruhe, but there are many obstacles to migratory fish in almost all of the tributaries. Only the barrages on the River Sieg have been modified with ramps and/or fish passages to allow fish movement. In the autumn of 1993, at least 14 salmon, mature to spawn, returned to the River Sieg, and in February 1994, newly hatched yolk sac alevins were found in the natural spawning grounds of the River Sieg and its tributary, the River Broel.

Cost: Expenditures for fish passages on the barrage weirs Iffezheim and Gambsheim and on weirs on the River Lahn: 12 million ECU =

about 24 million DM (EC subvention: 5 per cent; France, Germany, Rhineland–Palatinate: 95 per cent);

Overfishing

Problem: Intensive fishing of a stock in decline resulted in a further decrease of the stock, and, in some cases, in its disappearance. The reason for this was inadequate management and the absence of regulations. In the future, fishing may be permitted only to the extent that it will not endanger the stock of any species, that is, within a maximum sustainable yield.

Action: The restoration of the salmon by introducing young fish into the remaining intact habitats is considered to be a promising approach; if these hatchlings complete their downstream migration successfully, then they will return to the original habitat to spawn when mature.

Costs: Salmon restocking costs subsumed in the estimate for habitat restoration above.

SUCCESS OF 'SALMON 2000'

There are encouraging signs of ecological recovery in the Rhine ecosystem. Nearly all fish species formerly populating the Rhine have today returned to the river. Even specimens of migratory fish that disappeared several decades ago, such as sea trout, allice shad, and sea lamprey, have again been observed. In December 1990, news came of the first salmon landed in the Broel, a tributary of the Sieg. From its markings this salmon was recognized as having been released into the Broel in 1988. Specimens of other (locally) endangered species, such as schneider, barbel, nase, and flounder have also been sighted.

In addition to fish, invertebrates such as those living in or on the sediment and the banks of the river are part of the aquatic ecosystem. They are also highly sensitive indicators of the water quality. In 1990, 103 different species and groups were recorded. This number may be compared to the mere 27 species identified at the beginning of the 1970s, when the pollution of the Rhine had reached its peak.

In a recent publicity summary of progress on Salmon 2000, the ICPR says:

A lot remains to be done, we've only just started. It is firmly hoped that, by the year 2000, the target set in 1986 will have been reached: migrants—such as salmon—will have returned to the Rhine and its tributaries, and be self-supporting.

The Rhine will never be quite its old self. But it is still powerful, even though it is being limited by a canal corset. Some of its tributaries are nearly as beautiful as they were in former times or little needs to be done for them to reach that state.

Things cannot return to what they used to be. Today, it is not the old world otter that is waiting for salmon and sea trout, but man. Wise anglers know that these rare migrants depend on our protection. Osprey do not breed in the alluvial areas of the Rhine, but for some years, the number of breeding herons has increased. They too indicate that our old Father Rhine is recovering.

THE RIVER DANUBE

BASIC CHARACTERISTICS OF THE DANUBE RIVER BASIN

The Danube river has a total length of 2857 kilometres, a flood plain that covers 17,737 square kilometres, and a total drainage area of about 817,000 square kilometres (70 per cent of the total area of central Europe). The basin is home to 86 million people, 12 per cent of all Europeans. The river flows from alpine mountain streams, at some 3000 metres, to the delta at the Black Sea, where the mean flow is around 6550 cubic metres per second, with an extreme range of 1610 to 15,540 cubic metres per second.

The river may be divided into four sections (see Figure 2.3). The Upper Region flows from the source tributaries down to Bratislava, where the Rivers Brigach and Breg converge to form the Danube proper. Originally, the many tributaries contributing to the flow in this region added large amounts of sediment to the main river channel, but this has been greatly reduced by the construction of hydraulic works (dams and reservoirs).

The Middle Region is the largest section of the basin and extends from the confluence of the River Morava at Bratislava to the Iron Gate dams, where the Danube enters the flat plains. Much of the sediment carried from the upper region is deposited in the middle section, whilst the flow increases to around 5700 cubic metres per second as large tributaries join the main river from the north. At Moldova Veche (1048 kilometres from the source) the river enters a 117 kilometres long gorge section, between the Carpathians and Balkans, which has been extensively altered by reservoirs for hydropower and navigation.

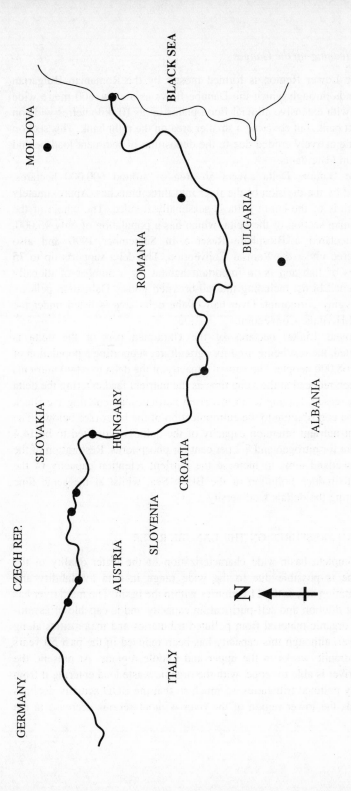

FIGURE 2.3: Map of the Danube River and Basin

The Lower Region is formed mostly by the Romanian–Bulgarian lowlands through which the Danube flows as a slow, 800 metre wide river, with extensive alluvial flood plains up to 10 kilometres wide on the left bank, but covering a smaller area on the right bank. This section is being actively eroded due to the deposition of sediment load behind the Iron Gate dams.

The Danube Delta covers an area of around 600,000 hectares. formed by the division of the river into three branches. Approximately two-thirds of the entire delta is seasonally flooded. The whole of the Romanian section of the delta, which has a population of only 15,000, was declared a Biosphere Reserve in September 1990 and also registered under the Ramsar Convention. The delta supports up to 75 species of fish and is an important habitat for a number of globally threatened birds, including the red-breasted goose, Dalmatian pelican, and pygmy cormorant. Over half of the delta area is listed under the World Heritage Convention.

Around 15,000 hectares of the Ukrainian part of the delta is protected, the rest being used for agriculture, supporting a population of about 68,000 people. The natural capacity of the delta to retain nutrients has been reduced at the same time as the nutrient load entering the delta has increased, leading to an increased nutrient load entering the Black Sea and contributing to the eutrophication of the latter (see below). The present nutrient retention capacity of the delta is estimated to be 14.4 per cent for nitrogen and 8.2 per cent for phosphorus. Restoration of the delta wetland aims to increase the nutrient retention capacity of the delta to reduce pollution of the Black Sea, whilst at the same time protecting the delta's biodiversity.

HUMAN PRESSURES ON THE DANUBE RIVER

No complete basin-wide characterization of the water quality of the Danube is possible due to the wide range in data availability and presentation across the 11 countries within the basin. The main river has a great dilution and self-purification capacity and is capable of assimilating organic material from polluted tributaries and major cities along the river, although this capacity has been reduced in the past 50 years by hydraulic works in the upper and middle regions. At present, the main river is able to 'cope' with the organic waste load entering it from heavily polluted tributaries, so much so that the BOD actually declines towards the lower region of the river without serious decrease in the oxygen content. There is thus great potential for significant saving in the need for waste water treatment if the natural self-purification capacity of the river is maintained by not overloading it with organic and inorganic waste. The quality of the Danube river water is, in fact, improving at present as a result of the downturn in heavy industrial economic activity following the political changes taking place in several countries within the basin.

The human pressures on the Danube river are similar to those affecting other major rivers in Europe and are summarized in Table 2.4.

Eutrophication represents a major threat to the ecological function of a large number of slow moving and still water bodies, not only within the Danube River channel and its tributaries, but also to the Black Sea into which the Danube drains. Eutrophication is not, at present, a serious problem to the main free-flowing river where dissolved oxygen levels are higher. The nitrate level in the main Danube is below current EU standards for drinking water extraction (Guide Level: 25 milligrams per litre; Maximum Acceptable Concentration: 50 milligrams per litre), and is still considered a 'safe' raw water source by this criterion. However, levels of nitrates have increased four to five-fold over the past 30 years and further increases would alter this situation, especially if fertilizer application rates increase with economic recovery. A considerable proportion of the water supply of the basin countries is still derived from bank-filtered water (for example 5.5 million cubic metres per day for Hungary alone). However, there are signs that the natural purification capacity of this simple system is being stressed by pollution (for example complaints about the quality of the 150,000 cubic metres per day Nussdorf water works, which is now used only as a reserve supply for Vienna).

Contamination of ground water resources by nitrate is a more serious problem than nitrate pollution condition of the river itself, because millions of people within the Danube basin are still dependent upon shallow hand-dug wells for their drinking water supply. Levels of nitrate and phosphorus, as well as chloride, sulphate, ammonia, and phenol, often exceed legal limits. Once in the ground water supply, removal or treatment is very expensive if not impossible.

The main cause of eutrophication is the agricultural sector, which contributes an average of 50 per cent of the total load of nitrogen and phosphorus to the Danube. Although total applications are lower in the Danube basin than in much of western Europe, over-application and improper application of fertilizer in 'sensitive areas' cause excessive

TABLE 2.4
Human Pressure on the River Danube

Water use problem	Drinking water supply	Fisheries	Industry	Irrigation	Recreation
Nutrient load and eutrophication	Nitrate contamination of ground water	Loss of sensitive species	Increased cost of treatment and reduction in some uses, e.g. cooling	Possible positive benefit	Loss of aesthetic value; reduction in recreation benefits
Heterotrophic growth and oxygen depletion	Surface water unfit for consumption; reduced infiltration to groundwater	Direct toxic effect of ammonia; loss of species if oxygen level drops below critical minimum	Increased cost of water treatment	Damage to modern irrigation equipment	Reduction in recreation benefits
Microbial pollution	Contamination of surface and ground water supply; cost of treatment	Toxic 'blooms'	Increased cost of treatment for some uses, e.g. food processing	Water unfit for some crops, e.g. raw salad species.	Reduction in recreation benefits; health treatment costs
Hazardous substances, including oils	Contamination of surface and ground water supply; cost of treatment; long-term contamination of sediments	Loss of sensitive species	Increased cost of treatment	Water toxic to crops or to man	Reduction in recreation benefits; health treatment costs
Competition for available water	Reduced or intermittent supply; increased prices	Loss of ecological habitat, especially spawning grounds	Reduced or intermittent supply	Reduced supply during growing season	Reduction in recreation benefits

Source: Strategic Action Plan for the Danube River Basin 1995–2005, UNDP.

drainage into waterways and ground water. Livestock practices in some Danube countries make a major contribution to eutrophication, for example in Romania the nutrient load from livestock is equivalent to that from a population of 25 million people. Recently, there has been a sharp decline in fertilizer input into agriculture in the region due to the economic transition (for example down from 268 kilograms per hectare per year of N/P/K in 1988 in Romania to 37 kilograms per hectare per year in 1991).

Human populations contribute about 25 per cent of the nutrient load of the Danube as the result of the absence, or poor working condition, of waste water treatment along the Danube below Austria. Although most major cities on the river have biological waste water treatment, the operational efficiency of this plant is often much lower than its original specification and some cities still discharge untreated sewage directly into the river (for example 75 per cent of sewage from Bucharest, the capital city of Romania).

The remaining 25–30 per cent of nutrient input arises from industrial sources, either directly as run-off from the phosphor–gypsum stores of fertilizer factories, or indirectly via atmospheric deposition.

Heterotrophic growth, resulting from carbon-based (organic) material discharged into the Danube, provides a biochemical substrate for many aquatic organisms which use dissolved oxygen in the water to oxidize the organic materials, thus reducing the suitability of the water for other species. The tributaries of the Danube, especially those with industrial plants discharging large volumes of organic waste (for example the sugar factory at Pleven on the River Vlit in Bulgaria) are more affected by this problem than is the main river, which has a large dilution and oxygen mixing capacity.

Microbial contamination by pathogenic bacteria, viruses, and protozoa is an important and growing problem in the river basin, although an adequate set of monitoring data does not yet exist. Faecal coliform counts for a few locations in the Danube basin have been found to exceed the EU guide value for bathing of 1 per millilitre and also the limit value of 20 per millilitre; examples collected in 1988 are given in Table 2.5.

The EU recommends that viruses should not be detectable in any water bodies intended for bathing, yet viruses have been found widely within the basin, including water samples from Austria and Germany, which have sophisticated water treatment works. This indicates that storm-water run-off and diffuse inputs from unsewered sources may be important contributors to the problem. Modern mechanical and biological sewerage treatment should remove 50–90 per cent of bacteria from waste water, but have less effect on viruses, but the inadequate capacity and lack of maintenance of existing plants in many parts of the basin result in a sub-standard performance, due to frequent bypassing or short retention times.

TABLE 2.5

Microbial Contamination at Sample Sites on the River Danube

Country	Location	Faecal coliform count (per ml)
Bulgaria	Nikopol	3
	Novo Selo	9
	Silistra	13
	Vidin	0
Hungary	Baja	13
	Visegrad	300
Romania	Galatz	20
Slovakia	Bratislava	40
Ukraine	Ismail	40
	Vilkovo	30

Source: Strategic Action Plan for the Danube River Basin, 1995–2005, UNDP.

It is likely that health problems arising from water-borne microbial diseases are still only a localized problem within the river basin, depending upon local conditions such as the exact nature of the microbial sources, the filtering properties of the local soils and the method of drinking water supply. The establishment of a more effective monitoring system for microbial contamination is a prerequisite before any control measures can be undertaken.

Contamination by hazardous substances within the Danube basin is so poorly documented that it is almost impossible to make more than generalized statements about the problem. A survey of all major industrial point sources of hazardous chemicals has yet to be completed, whilst virtually no capacity exists for monitoring (or containing) accidental spillages—even the oil-spills which occur almost daily.

The hazardous substances of particular concern are pesticides, ammonia, PCBs, PAHs, and heavy metals. Current dissolved levels of most of these substances are not excessive, but the build up of persistent compounds in riverbed sediments represents a major mid-term threat. A 1991 survey revealed that sediment at 23 out of 55 sites would be classified as hazardous waste under Dutch regulations.

With the exception of ammonia (from agriculture), the major source of hazardous pollutants is mining and other major industrial plants plus oil discharged from river shipping and old and poorly maintained pipelines. Once major point sources have been identified, it requires the transfer of existing (but expensive) technology in order to prevent these chemicals from entering the main river system, because it is even more expensive if not impossible to remove these pollutants via existing municipal waste water treatment plants.

Competition for available water is already a serious local problem in some parts of the Danube basin, notably on the left bank tributaries in Hungary and the tributaries of Romania and Bulgaria. Existing shortages have been exacerbated by a 10 year drought, with water flow in 1993 being the lowest for 100 years. In spite of the construction of numerous multi-purpose dams and reservoirs (over 400 in Romania alone), the inefficient use of water for irrigation and industry in Central European Countries or CECs resulted in serious drinking water shortages, even in major cities like Sofia.

THE RELATIONSHIP BETWEEN THE RIVER DANUBE AND THE BLACK SEA

The Danube is the major freshwater river and provides the major nutrient input flowing into the Black Sea. The nutrient load of nitrogen and phosphorus entering the Black Sea from the Danube has increased ninefold over the period 1960 to 1990. The flow-corrected figures for 1991 indicated that 540,000 tonnes of nitrogen and 45,000–50,000 tonnes of phosphorus were carried into the Black Sea. There are large discrepancies in the estimates of the nutrient inputs and outputs within the basin, but it is not known to what extent these reflect poor statistical data or the ability of the river wetlands to retain nutrients.

The heavy nutrient load carried by the Danube into the Black Sea is a much more serious ecological threat to the latter than to the Danube itself. The Black Sea has been described as being 'within an inch of falling into a coma'; the waters are highly stratified, and the chemocline (depth at which oxygen is severely reduced) has risen from 170 metres to 110 metres, reducing the productivity of fisheries.

The World Bank, through the Global Environment Facility (GEF) has recently approved an advance to provide technical assistance and equipment to the Romanian government to prepare a US $ 4.2 million Danube Delta Biodiversity project. This will support local authorities in their efforts to manage the delta as an ecological system, including environmentally sustainable economic activities (fisheries, small-scale agriculture, tourism). This project will supplement two other regional GEF projects, Environmental Programme for the Danube River Basin (see below), and the Black Sea Environmental Programme.

POLICY AND ACTION TO CONTROL ENVIRONMENTAL PROBLEMS IN THE DANUBE BASIN

The first major international agreement to address the declining environmental conditions in the Danube basin was the Bucharest Declaration, signed in 1985 by the (then) nine countries through which the Danube flows. The Bucharest Declaration is an agreement to cooperate on water management and pollution control, but real progress on these matters has been stymied by socio-economic and political problems in the region. However, during this period, renewed efforts have been made to regenerate the international collaboration that will be required to address environmental problems in the region.

THE ENVIRONMENTAL PROGRAMME FOR THE DANUBE RIVER BASIN

The United Nations Development Programme (UNDP) GEF Environmental Programme for the Danube River basin was conceived in Sofia in September 1991 and started in 1992, following agreement between its parties on a Work Plan and establishment of a Task Force. The Programme provides for joint action over a period of three years to commence the task of improving environmental management in the basin. The drafting of the Strategic Action Plan (SAP) for the Danube River Basin (1995–2005) was one of the main objectives for the Task Force.

The aim of the GEF project was to create the framework for a long-term solution to the problem of pollution in the Danube river basin. The project will form the scientific, institutional, and strategic basis for an action programme for Phase 1 of the Environmental Programme for the Danube River Basin by:[5]

- collecting data and establishing national and regional databases and information systems;
- providing technical assistance to participating countries' governments in identifying key problems and developing overviews of the existing situation;
- establishing networks of information exchange among like groups in participating countries, and providing training and institutional strengthening;
- developing a strategic action plan for addressing water pollution in the Danube basin; and
- preparing a series of feasibility studies for high priority investments for local and international funding.

Production of the SAP was supported financially by the UNDP Regional GEF project for Environmental Management in the Danube River Basin (RER/91/G31/A/1G/31). The GEF project was financed by US $ 8.5 million of UNDP/World Bank/United Nations Environmental Program or UNEP funds. These funds were to be matched in kind by Government inputs from the 11 countries situated within the Basin (Austria, Bulgaria, Croatia, the Czech Republic, Germany, Hungary, Moldovia, Romania, the Slovak Republic, Slovenia, and the Ukraine).

Additional funding for the programme has been provided by the European Commission PHARE Regional Programme, the European Bank for Reconstruction and Development, Austria, the Netherlands, the USA, and the Barbara Gauntlett Foundation, whilst Danube countries and the European Commission (General budget) have contributed to staff, accommodation, and operating costs for the Programme Co-ordinating Unit (PCU). Danube countries have supported the programme through the provision of national expertise, country information, and, where possible, facilities for meetings and workshops.

The major output from the Task Force has been the production of the SAP, which was approved by the Task Force on 28 October 1994 in Bled, and was endorsed by the Environment/Water Ministers of the Danube countries and the Member of the European Commission

---

[5]*UNDP GEF Regional Environmental Management in the Danube River Basin Project Document.*

responsible for the Environment in the Ministerial Declaration of Bucharest on 6 December 1994. The SAP is the main practical document of the continuing Environmental Programme for the Danube River Basin and is summarized below.

THE STRATEGIC ACTION PLAN FOR THE DANUBE RIVER BASIN (1995–2005)

The Action Plan has four major goals:

- to reduce the negative impacts of activities in the Danube river basin and on riverine ecosystems and the Black Sea;
- to maintain and improve the availability and quality of water in the Danube river basin;
- to establish control of hazards from accidental spills; and
- to develop regional water management co-operation.

The SAP will be achieving these goals via a series of strategic directions covering key sectors and policies, including:

- phased expansion of sewerage and municipal waste water treatment capacity;
- reduction of discharges from industry;
- reduction of emissions from agriculture;
- conservation, restoration, and management of the wetland and flood plain areas of the tributaries and main stream of the Danube River Basin;
- integrated water management;
- environmentally sound sectoral policies;
- control of risk from accidents; and
- investments.

The SAP identifies the following list of common targets in the short, medium, and long term:

Short term targets:

- Elaboration of National Action Plans (NAPs) for implementation of the SAP;
- Completion of integrated tributary river basin plans and revised water allocation and water use permits;

- Completion of wetland inventory, conservation, and management programmes;
- Adoption of consistent water quality objectives and criteria for all Danube tributaries and the main river;
- Adoption of emission limits for fertilizer plants;
- Adoption of emission limits based on Best Available Technology (BAT) for new industrial sources and adoption of emission limits for livestock enterprises;
- Completion of a comprehensive system of information on the state of the river environment;
- National assessment of critical loads and load reduction targets for the higher priority river reaches;
- Evaluation of the critical load of nutrients from the Danube into the Black Sea;
- Completion of effective and comprehensive monitoring, warning, and laboratory systems, including systems for exchange of information;
- Development of technical and management capacity building programmes for all actors and sectors.

Medium term targets:

- Completion of pollution emission inventory;
- Adoption and implementation of hazardous substance control legislation, including transport;
- Introduction of regulations for fertilizer storage, handling, and application;
- Preparation of waste water and sewerage investment priorities for cities, rural towns, and villages;
- Completion of projects on conservation and restoration of priority wetlands;
- Investments in high priority sewerage and municipal waste water treatment capacity extensions;
- Introduction of environmentally sound agriculture policy reforms;
- Demonstration of Best Environmental Practice (BEP) or use of fertilizers, pesticides, and other agrochemicals in agriculture;

- Completion of pilot and demonstration projects for manure handling, storage, disposal, and application;
- Introduction of phosphate-free detergents and ban on phosphate-containing detergents;
- Phased application of emission limits and incorporation of emission limits into permits for industries affecting critical reaches and sites.

Long term targets

- Completion of construction of municipal and industrial waste water treatment plants;
- Change to sustainable agricultural practices;
- Restoration of the natural purification capacity of the Danube and its tributaries.

Recognizing that achieving all these targets will probably require more than the full 10 year time frame of the project, the SAP places great emphasis on the prioritization of actions and stresses the urgency with which national governments and organizations should:

- Prepare a list of costed immediate priority actions addressing major health, ecological, and economic risks; indicating who should undertake the priority actions, how the financing will be obtained, and the time scale;
- Identify a mechanism and the necessary resources to support the development and production of the National Action Plans;
- Assess the risks and implications for public health and the protection of ecosystems if the short and medium term targets in the SAP are not met.

No total estimate of the costs for the SAP have been prepared to date. However, it is recognized that they are large and that although the international community can help the Danube countries in the financing of priority actions, long term financing will have to be met primarily from within the countries themselves. It is clear that there is a significant resource gap between the proposed actions and the available funding. To address this, the following actions should be taken:

- Develop a financing plan (domestic and international) in order to identify what funds are available and what funds are needed to meet the most urgent and short term priorities;
- Initiate international discussions in governments involving the Ministry of Finance and all relevant ministries to develop a financial plan for dealing with the most urgent and short term priorities;
- Earmark funds for transferring training and know-how to the riparian governments on raising funds for environmental initiatives;
- Initiate mechanisms that can make loans for environmental improvements more realistic, attractive, and affordable for riparian countries;
- Provide a list of initiatives which have resulted from the pre-investment studies of projects which are being considered for development and/or funding by the international financial institutions.

PRIORITY ACTIVITIES AND COSTS

Hot spots

The Danube countries have drawn up a comprehensive list of 'hot spots', defined as: 'local land areas, stretches of surface water or specific aquifers which are subject to excessive pollution from an identifiable source and which require particular actions to prevent or reduce the degradation caused'. The SAP contains a table with 178 priority projects, of which 102 are flagged as being of the highest priority. These projects are distributed across the 11 Danube countries as shown in Table 2.6. At the time of production of the SAP, only 44 of these projects had cost estimates of any kind.

From Table 2.6, it may be calculated that the total cost of projects proposed under the SAP would be well over US $ 4 billion if the estimated costs in the table are representative of the other projects. However, over 50 per cent of the estimated costs arise from a single project to upgrade the sewerage treatment capacity of Budapest (Hungary). Removing this project brings the average estimated cost per project down to around US $ 11 million and the total cost of the other projects proposed under the SAP to just under US $ 2 billion (excluding the Budapest project).

TABLE 2.6

Hot Spot Projects within the Environmental Programme for the Danube River Basin

Country	Number of projects proposed	Highest priority projects	Costed projects	Total cost (US$ million)
Austria (+)	0	0	0	–
Bulgaria	29	14	11	72.015
Croatia	22	11	2	7.558*
Czech Republic	10	5	8	70.160
Germany (+)	0	0	0	–
Hungary	29	17	20	816.000†
Moldova	7	7	0	–
Romania	43	29	0	–
Slovakia	22	6	0	–
Slovenia	13	9	3	43.500
Ukraine	4	3	0	–
Total	*179*	*101*	*44*	*1009.233*

(+) No hotspots are listed for Germany and Austria, illustrating the success of these countries' better established water quality legislation and investment in municipal and industrial wastewater treatment.

*Originally costed as ECU 4.031 million

†US$ 540 million of this total is for the MWWTP for Budapest.

Wetland Restoration

In addition to the priority 'hot spot' projects described above, the Danube SAP proposes an extensive programme of wetland restoration along the entire river length. A total of 59 projects have been proposed, of which 34 have estimated costs totalling ECU 25.412 million (see Table 2.7).

TABLE 2.7
Wetland Restoration Projects within the Environmental
Programme for the Danube River Basin

Country	Number of projects proposed	Costed projects	Total cost (ECU million)
Austria	5	1	2.500*
Bulgaria	5	5	0.540
Croatia	4	3	12.100
Czech Republic	8	1	0.775
Germany	8	1	6.545†
Hungary	13	12	2.210‡
Moldova	1	0	–
Romania	2	0	–
Slovakia	6	6	1.127‡
Solvenia	5	5	0.390
Ukraine	2	0	–
Total	59	34	25.412

*Estimate of ECU 2–3 million, plus ECU 3–6 million annual operating costs;
†Original estimate DM 10–15 million plus loss of hydropower capacity
 (DM 1.91 = ECU 1).
‡Some projects estimated in US$ (converted at US$ 1.248 = ECU 1)

The projected cost of the entire programme might therefore be estimated at ECU 44.09 million but, excluding one single project from Croatia estimated at ECU 10 million, the projected cost becomes ECU 27.09 million for the remaining projects.

Economic Values of the Danube Flood plain

To date, there has been no attempt to calculate a Total Economic Value (TEV) for the Danube river. Recently, an estimate of the economic value of the Danube flood plains in their present semi-natural condition has been made, which arrived at a total annual value of ECU 650 million (Gren, Groth, and Sylven 1995). This estimate involved some 'heroic' simplifications and should be interpreted with considerable caution, but the methodology used can provide useful insights into how to approach these problems.

The major functions of the present semi-natural flood plains of the River Danube are:

- Microclimate control;
- Flood control;
- Self-purification following inputs of human waste;
- Nutrient sink;
- Provision of drinking water via bank filtration and ground water recharge;
- Ecological habitat for wildlife; and
- Recreational use.

For valuation, these functions can be grouped together under the following categories:

- Inputs for production of marketed good and services, for example timber, fisheries, fodder;
- Production of eco-technologies, for example flood plain forests as carbon sinks, wetlands as nutrient and pesticide sinks;
- Consumption goods, for example hunting and other recreation activities.

Not all of these functions can be valued accurately and, in the Danube Basin, there are few data upon which to base any estimate of value. Using estimates of value from studies in other locations, Gren *et al.* arrive at the following overall values for the Danube flood plain (see Table 2.8).

The average values for input resource, recreation, and nutrient sink functions across the Danube Basin are ECU 61, 101, and 212 per hectare respectively. These values vary with economic market strength of the different countries within the basin, and with market prices (and hence values) being higher in Germany and Austria than in the other countries.

The loss of the functions outlined above would (probably) be permanently lost if the flood plain were converted to other uses incompatible with existing functions. It is therefore necessary to calculate the net present value (NPV) of the existing functions, not just the annual

TABLE 2.8

Estimated Value for Different Countries in the Danube River Basin

Country	Flood plain area (ha)	Input resource value (ECU/yr)	Recreation Value (ECU/yr)	Nutrient sink value (ECU/yr)	Total value (ECU/yr)
Germany	45,662	5,022,820	8,219,160	9,680,344	22,922,324
Austria	27,500	3,025,000	4,950,000	5,830,000	13,805,000
Slovakia	5000	295,000	485,000	1,060,000	1,840,000
Hungary	51,553	3,041,627	5,000,641	10,929,236	18,971,504
Croatia	350,000	20,650,000	33,950,000	74,200,000	128,800,000
Bulgaria	80,000	4,720,000	7,760,000	16,960,000	29,440,000
Romania	1,028,000	60,652,000	99,716,000	217,936,000	378,304,000
Ukraine	150,000	8,850,000	14,550,000	31,800,000	55,200,000
Total	1,737,715	106,256,447	174,630,801	368,395,580	649,282,828

Source: Gren, Inge-Marie, Klaus-Henning Groth and Magnus Sylven (1995).

value. Applying a discount rate of 5 per cent, Gren *et al.* derive a NPV of ECU 7480 per hectare, thus the calculation above would rise considerably if future benefit streams were included.

CONCLUSIONS

This chapter has reviewed the river clean-up projects from the Thames, the Rhine and the Danube and has tried to draw out the lessons for the Ganga Action Plan. A number of features stand out from the three river schemes reviewed.

First, they all have a long history. Action has been taken over periods of 20 years and more. It would be unreasonable for a developing country to expect to achieve in 10 years the same as what has been achieved by countries with many more resources over 20 years.

Second, the costs involved have been enormous. Although exact figures are impossible to come by, the orders of magnitude are clear. The Thames investments have amounted to over £100 million. in the period 1950–80. At 1995 exchange rates this would amount to over Rs 5000 million. By contrast the Ganga project, which runs over 2500 kilometres compared to 245 kilometres for the Thames has an investment budget for GAP I and GAP II of Rs 11,200 million. In this respect the GAP level of expenditure does not seem excessive, although GAP has yet to achieve water quality goals comparable to the Thames. The Rhine programme was even larger. For a river 1320 kilometres long, the expenditures have been around DM 100 billion between 1965 and 1989, primarily for the construction of wastewater treatment plants. At 1995 exchange rates this amounts to about Rs 1940 billion, or 176 times the GAP. Again the levels of clean-up required, and achieved, are not directly comparable, but it helps put the GAP in perspective. Finally the Danube programme for a river of similar length to the Ganga started out more modestly. Much is still to be done and a prioritized strategic action plan has been drawn up with an expected cost of $4 billion, or around Rs 125 billion. This is a plan that focuses on hot spots and makes its selection of projects, at least in part, on a cost-benefit priority basis.

Third, and following on from the last point, the selection of actions has not been determined by the comparison of costs and benefits. A system of clean rivers is taken politically as something which all civilized nations should aim for, and whatever costs are required to achieve those goals are considered as reasonable. Of course, the programmes have focused on interventions that are 'cost-effective' that is those that get a given improvement in quality at lowest cost, but even this has not been pursued as strictly as it might. Certainly, the selection of water quality improvements on the basis of a comparison of benefits and costs is a novel approach in river clean-up. Only recently, with the Danube, has this been pursued, and even there it is in its infancy. Given the high costs involved in such programmes, it is surprising that more has not been done in a cost–benefit framework. The present study is therefore a contribution to the general literature on benefit estimation for river projects, as well as an evaluation of the GAP.

3

Methodology

INTRODUCTION

The method used in this study is one of extended social cost–benefit analysis, where project costs and marketable and non-marketable benefits are quantified in monetary terms to the maximum extent possible. Some of these may not involve actual payments in cash, or even in kind. For example, the benefit of a clean Ganges to an Indian who does not live near the river, and who does not visit it, will have no cash payments involved. But it is, nevertheless, a value and should be measured and included in the cost–benefit analysis. For an environmentally oriented project like the GAP, many benefits are in this form. Techniques for their valuation have advanced considerably and methods are now available that allow for a wide range of environmental values to be estimated.[1]

Section 3.2 discusses the problem of identification of benefits and costs of the GAP. Two methods of classification of benefits are presented. One classification is between the marketable and non-marketable benefits and the other is between user and non-user benefits. Section 3.3 describes the problem of measurement of benefits and costs and also provides a brief review of methods of valuation of non-marketable benefits. Finally, Section 3.4 deals with the problem of measuring social costs and benefits after taking into account the differences between the market prices and shadow prices, the income distribution effects, and the intertemporal equity issues in resource use.

IDENTIFICATION OF BENEFITS AND COSTS

There can be a variety of benefits from the improvement of quality of

[1]For a review of these methods, see Mitchell and Carson (1989); Freeman (1993); Oakridge National Laboratories or ONL/Resources for Future or RFF (1994): Markandya *et al.* (1999).

water in a freshwater body like the Ganges. These can be classified as user benefits and non-user benefits. Table 3.1 from Mitchell and Carson (1989) provides details about the categorization of benefits from the improved freshwater. The user benefits can be classified as marketable benefits: agriculture, fisheries, navigation, process water for industry, and water for household uses; and non-marketable environmental benefits: recreation, and waste disposal services for industry and households. There are non-user benefits from vicarious consumption (deriving pleasure by feeling that others get benefits from a preserved resource), stewardship, and bequest benefits. The non-user benefits are also categorized as existence value benefits.

Benefits from cleaning the Ganges or the Ganga Action Plan (GAP) can be broadly divided into two categories: first, those that accrue to people who stay near the river or visit the river for pilgrimages or for tourism, and second, those accruing to people who do not stay near the river but enjoy certain benefits from simply knowing that the Ganges is now a cleaner river. Benefits for the first category of people, who use the river directly, may include bathing, washing, drinking, and fishing in cleaner water. The GAP projects, especially the sewage treatment plants, provide irrigation and fertilizer benefits to farmers. Also, there are employment benefits from various GAP projects to unskilled labourers who are otherwise unemployed or underemployed in the Indian economy. There are also benefits in the form of savings in the cost to the water supplying undertakings along the Ganges due to the GAP.

Benefits to the second category of people, sometimes called indirect-user benefits or non-user benefits, are less apparent at first sight. But, on reflection, it is possible to identify several reasons why a cleaner Ganges may give a sense of satisfaction to people living as far away from the river as Madras (Chennai), Trivandrum (Thiruvananthapuram), Hyderabad, or Baroda.

First, one may feel a general sense of satisfaction from just knowing that a major river in the country is now cleaner than it was earlier, whether this stems from ecological considerations or from the point of view of pure scenic beauty. Or there may be other specific reasons for this feeling of increased well-being.

Second, there could be some perceived benefits from knowing that a river of major religious significance is now cleaner. The Ganges epitomises purity in Hindu mythology and although there is the prevalent, almost tautological, belief that a 'pure' river cannot be polluted, most people who go to have a 'holy dip' are able to perceive a benefit in having a less polluted river to bathe in.

TABLE 3.1

Classification of the Benefits from an Improvement
in Freshwater Quality

Benefit class	Benefit category	Benefit subcategory (examples)
Use	In-stream	(1) Recreational (water skiing, fishing, swimming, and boating)
		(2) Commerical (fishing, navigation)
	Withdrawal	(1) Agriculture (irrigation)
		(2) Industrial/commerical (process treatment, waste disposal)
	Aesthetic	(1) Enhanced near-water recreation (hiking, picnicking, photography)
		(2) Enhanced routine viewing (commuting, office/home views)
	Ecosystem	(1) Enhanced recreation support (duck hunting)
		(2) Enhanced general ecosystem support (food chain)
Existence	Vicarious	(1) Significant others (relatives, close friends)
	Consumptio	(2) Diffuse others (general public)
	Stewardship	(1) Inherent (preserving remote wetlands)
		(2) Bequest (family, future generations)

Source: Mitchell and Carson (1989), Figure 3.1, p. 61.

There is also the extended-welfare basis of feeling a sense of satisfaction from knowing that fellow citizens living near the river have a better quality of life with a cleaner river. Those living near the river have a better quality of life in the sense that they are less prone to water-borne diseases and infections, and from the point of view that they will have cleaner waters to irrigate crops and to catch fish from.

From an ecological point of view, there can be a sense of satisfaction from knowing that biodiversity has improved, given that cleaner waters mean that more species of flora and fauna can exist in the river.

There is the pleasure of knowing that in the future the water will be cleaner for recreational uses like swimming, boating, or picnicking on its banks, and that these benefits and the simple one of simply beholding a clean river are already being reaped by friends and relatives who may live near the river.

Finally, from the perspective of future generations, there is a sense of enhanced well-being from knowing that our children and their children will inherit a cleaner river resource in the future.

These, in brief, are the non-user benefits that can be derived from a cleaner Ganges, and by extension, from any plan to make the river cleaner, such as the Ganga Action Plan.

As will become clear, the list of potential non-user benefits identified from the Ganga Action Plan includes all the possible sources of benefits listed under the 'Existence' category of benefits of the Mitchell and Carson typology. In addition, it details the benefits listed in the 'use' category of benefits, from an indirect-user point of view. For example, even someone who is not necessarily a resident or a visitor to the river can appreciate the ecosystem effects of cleaner water. The same is the case for the benefits to fishermen living on the banks of the river. The argument for their inclusion in the list of indirect benefits is simply that all these can engender a feeling of satisfaction among citizens of the country resident elsewhere in the country.[2]

The cost of cleaning the Ganges is borne by the government (the Union government and the state governments of Uttar Pradesh, Bihar, and West Bengal) and industry. While the various projects of the GAP deal with household-borne effluents, the industries located in the Gangetic basin have to control pollution to meet the prescribed pollution standards as per the environmental laws in India [for example the Water (Prevention and Control of Pollution) Act, 1974; the Air (Prevention and Control of Pollution) Act, 1981; and the Environment (Protection) Act, 1980]. Therefore, the cost of controlling industrial water pollution in the Gangetic basin is borne by industry.

GAP is a good example of how environmental federalism works in a federal country. The Ganga Directorate is a part of the Ministry of Environment and Forests (MoEF), and an apex body of GAP is located in Delhi, with its regional offices in Allahabad, Patna, and Calcutta in the states of Uttar Pradesh, Bihar, and West Bengal respectively. The

---

[2]This argument can, of course, be extended to Indians living in other countries, and also to citizens of other countries who may derive a sense of satisfaction from knowing that the Ganges is now cleaner than it was ten or more years ago.

GAP consists of schemes for the diversion and interception of industrial and household effluents, via the construction of sewage treatment plants, schemes of low cost sanitation, electric crematoria, schemes for river front development, and others. The Ganga Directorate, the various ministries of state governments like the Ministry of Irrigation and Public Works, and the departments of local municipal governments like Water Supply, Sewage and Sanitation are involved in the design and construction of these schemes. State government departments, in collaboration with regional offices of the Ganga Directorate and State Pollution Control Boards are responsible for the operation and maintenance of the GAP schemes. There is private sector participation in cleaning the Ganges, with many industrial units belonging to 17 highly water polluting industries in the Gangetic basin having effluent treatment plants to meet the water pollution standards fixed by the State Pollution Control Boards.

The decisions about the schemes, technologies, and investments of the GAP are taken by the Ganga Directorate. However, some inputs are provided by state government departments like Water Supply and Sewage Treatment Undertakings for designing of schemes such as the construction of sewage treatment plants. The regional offices of the Ganga Directorate and state government departments undertake execution, operation, and maintenance of the various schemes.

MEASURING BENEFITS AND COSTS

The marketable benefits of the GAP can be measured using market prices. The health benefits can be measured as health damages avoided by the users of river for bathing and drinking water due to the reduced level of pollution of the river. The market prices can be used to estimate the cost of potable water from alternative sources and to estimate the cost of illness from water-borne diseases. Chapter 7 describes the estimation of health benefits. Market prices can also be used to estimate benefits to fishermen from the increased fish production and to estimate the irrigation and fertilizer benefits to farmers. Chapters 9 and 10 deal with the estimation of fisheries and irrigation benefits. In the case of the investment costs and operation and maintenance costs of the GAP and the costs of water pollution abatement to industry, the market prices can be used. Chapter 4 describes the measurement of costs.

The measurement of benefits that do not have markets of exchange, and hence prices, poses problems. But these have recently been addressed successfully by a range of valuation methods developed in environmental economics. Of these, the contingent valuation method shows the most promise. But before detailing the method and its advantages, it is useful first to clarify the problem at hand.

There are several categories of benefits which do not pass through a market, and which, hence, do not have a market price to help measure them in money terms. In economic theory, these are called commodities whose markets are 'missing'. For example, although a clean river, with attendant riverine flora and fauna, generates aesthetic benefits to viewers (occasional or routine) there is no market to fix the price for these benefits—say in terms of an 'entry fee' charged to view the river for a certain period of time. As a rough parallel, zoos or nature reserves charge some entrance fee, which is the market price for the aesthetic pleasures that these facilities are able to provide.[3]

In the absence of markets for such benefits, therefore, the measurement of these benefits raises some difficulties. Methods to try and solve the measurement (or valuation) problem in the case of cost-and-benefit-streams from environmental resources (sometimes called 'service flows from resource-environmental systems' (Freeman, 1993: p. 6), for which markets do not exist, can be classified into two basic types: (1) those based on physical linkages and (2) those based on behavioural linkages (Smith and Krutilla, 1982).[4]

Valuation methods based on physical linkages essentially estimate physical effects (measured using dose-response functions or damage functions) and then value these physical effects using market prices (Mitchell and Carson, 1989: p. 74). Although these are frequently used in practice, and are preferred by natural scientists, there is little (welfare economic) theoretical basis for the use of these methods. From the point of view of economic theory the approach has several inadequacies (for example Mäler, 1974; Freeman, 1979) and at best they may be 'a first

---

[3]These 'prices', however, do not necessarily reflect the value of benefits that can be derived from the facilities themselves, since there are a variety of user and non-user benefits that such facilities provide that are not usually taken into account when setting these prices. Typically, they are based on the idea of covering costs rather than assessing value.

[4]From the point of view of formal economics, the value being estimated is the welfare measure of the benefits or costs of a change in environmental quality. These measures are either the compensating surplus or equivalent surplus, since the consumer equilibrium is quantity-constrained.

approximation to benefit measurement' (Mitchell and Carson, 1989: p. 75). In particular, they omit indirect-use benefit categories. If these are substantial, the resulting value of benefits will seriously be an underestimate of the value of total benefits. For this reason economists generally prefer valuation methods based on behavioural linkages, and only resort to damage function estimates when either time or resources do not permit the use of behavioural linkage methods. Damage function estimates are also sometimes computed to compare values of benefits across behavioural linkage methods and physical linkage methods. Typically, the latter should be a lower bound for the former. Valuation methods based on behavioural linkages basically try to assess values placed on the effects of changes in the provision of the commodity in question (in terms of either its quantity or its quality) by looking at the behaviour (and preferences) of affected individuals.

Behavioural linkage methods of valuing resource flows can be classified into four categories (Mitchell and Carson, 1989: p. 75; Freeman, 1993: p. 24), based on (a) how consumer preferences are revealed (observed market behaviour or responses to hypothetical markets) ; and (b) the type of behavioural linkage (direct or indirect linkage), as shown in Table 3.2.

TABLE 3.2

Behaviour-based Methods for Estimating Resource Values

	Direct behavioural linkage	*Indirect behavioural linkage*
Observed market behaviour	*Direct observed* Referenda Simulated markets Parallel private markets	*Indirect observed* Household production Hedonic pricing Actions of bureaucrats/ politicians
Responses to hypothetical markets	*Direct hypothetical* Contingent valuation allocation game with tax refund Spend more-same-less survey	*Indirect hypothetical* Contingent ranking priority evahuation technique Conjoint analysis question

Source: Mitchell and Carson (1989), p. 75.

The only method among these that: (1) can estimate *all* benefits— use benefits as well as existence benefits (to the extent to which they obtain); (2) has the ability to value goods not previously (or currently) available (since the market in question is hypothetical); and (3) can produce direct and unique estimates of the relevant ordinary and Hicksian inverse demand curves, however, is *contingent valuation.* This method presents respondents with a detailed scenario of the hypothetical market and of the public good in question (including the nature of the benefits it would confer on the respondent), and then directly asks respondents to state how much money they would be willing to give up (pay) in order to enjoy the benefits they would derive if that public good were provided. The general features of this method, and the specific methodology adopted in the valuation of non-user and user benefits from the Ganga Action Plan, are spelt out in Chapters 6 and 7.

The philosophy underlying the valuation is based on individual preferences, which are expressed through the willingness to pay (WTP) for something that improves individual welfare, and the willingness to accept (WTA) payment for something that reduces individual welfare. The total value of environmental impacts is taken as the sum of the WTP or WTA of the individuals comprising it. Thus, no special weight is given to any particular group. The WTP/WTA numbers can be expressed for both user and non-user values.

Although the valuation of environmental impacts using money values is widespread and growing, there are still many people who find the idea strange at best, and distasteful and unacceptable at worst. Given the central role being played by monetary valuation in this exercise, a justification of the method is warranted.

One objection to the use of WTP is that it is 'income constrained'. Since you cannot pay for what you do not have, a poorer person's WTP is less than that of a richer person, other things being equal. This occurs most forcefully in connection with the valuation of a statistical life (VOSL) where the WTP to avoid an increase in the risk of death is measured in terms of a VOSL. In general one would expect the VOSL for a poor person to be less than that of a rich person. But this is no more or less objectionable than saying that a rich person can and does spend more on health protection than a poor person; or that individuals of higher social status and wealth live longer on average than persons of lower status. The basic inequalities in society result in environmental values varying with income. One may object to these inequalities and make a strong case to change them. If social preferences are clear in this regard, the correct way to adjust the figures is to estimate the WTP and then apply 'distribution' weights.

SHADOW PRICES AND SOCIAL BENEFITS AND COSTS OF THE GAP

Having estimated the benefits and costs using the above methods, these estimates have to be corrected for differences between the market prices and shadow prices. In the literature on social cost-benefit analysis, there are now two methodologies that use shadow prices. They are known as UNIDO (Dasgupta, Sen, and Marglin, 1972) and OECD (Little and Mirrlees, 1974) methodologies.[5] The UNIDO methodology prescribes the estimation of the benefit and cost flows at market prices to start with, and then corrects for shadow prices for imperfections in the specific input and output markets. It suggests that imperfections abound in three important markets: capital, unskilled labour, and foreign exchange in the developing countries and the costs of these inputs have to be estimated at shadow prices instead of at their market prices in appraising investment projects. It also prescribes detailed methodologies to estimate the shadow prices of capital, unskilled labour, and foreign exchange. The OECD methodology prescribes world prices (free on board or f.o.b. prices for exports; and cost, insurance and freight inclusive or c.i.f. prices for imports) as shadow prices of commodities. For non-tradable commodities like power, transport, irrigation, and unskilled labour, methodologies are suggested to estimate the standard conversion factors or accounting ratios for finding out their values in foreign exchange. In case a country enjoys monopoly power in trade with respect to some commodities (its export and import decisions influencing commodity world prices), the marginal export revenue and marginal import cost in foreign exchange are prescribed as shadow prices.[6]

The UNIDO method is followed here in making the corrections for shadow prices in the estimation of the benefits and costs of GAP at market prices. The corrections are made only for shadow prices of capital and unskilled labour. The correction for the shadow price of foreign exchange is not attempted since the GAP does not have significant effects on the supply and demand for foreign exchange in the Indian economy. Many empirical studies on the social cost benefit analysis in

---

[5]UNIDO: United Nations Industrial Development Organization,

   OECD: Organization for Economic Co-operation and Development.

[6]Also see Squire and van der Tak (1975) and Ray (1984) for the discussion of some of these issues.

India suggest that the social time preference rate in India is lower than the rate of return on capital (the level of savings is suboptimal) and therefore there is a social premium on investment as high as 80 per cent (Murty *et al.*, 1992). This means the social opportunity cost of a rupee of investment in a public sector project like the GAP is Rs 1.80 in India. Also in the case of unskilled labour, there is disguised or underemployment in the rural or agricultural sector, so that the wage rate paid by the industrial projects is higher than the social opportunity cost of employing unskilled labour.

An extended cost-benefit analysis should examine not only the overall costs and benefits, but also look at who gains and who loses. The agents receiving benefits from GAP can be identified as users, non-users, unskilled labour, fishermen, farmers, and those gaining from improved water supply, while the agents who incur costs are industries and government. The distribution of benefits and costs of GAP among these agents is important to evaluate the effect of the project on the income distribution in the economy. Non-users and industry owners belong to high-income groups while the users, fishermen, and farmers belong to low income groups in the Indian economy. If the government has income distribution preferences, higher social values are associated with gains to lower income groups than to higher income groups. Equally, higher social costs are associated with losses for lower income groups than for higher income groups. There are various methods of estimating the social values or distribution weights attributable to incomes of people belonging to different income classes (Weisbrod, 1968; Dasgupta, Sen, and Marglin: 1972). The way in which this has been taken into account is by estimating the value of one rupee to a person as a function of his or her income. The method of calculating the weights is set out in Chapter 12.

Finally, the rate of discount determining the relative prices of present and future benefits of an investment project is an important national parameter in the investment project appraisal. It is an important issue in the context of the use of environment resources because many of the environmental benefits of present actions will occur many years from now and the higher the discount rate, the lower the value that will be attached to these damages. Furthermore, since the current use of environmental resources can affect the living standards of future generations, the discount rate used in the evaluation of preservation projects may reflect the policy makers' concern for intergenerational equity in resource use. It is argued that the social time preference rate is lower than the market rate of interest because of the presence of externalities in capital accumulation which are normally not taken into account by the individuals while making their saving–consumption decisions in the market (Markandya and Pearce, 1991) and intergenerational equity issues in resource use.

While there may be some merit in using a lower rate of discount for all public sector projects, it would be inappropriate to use a lower rate just for this one project. Since the rate used to decide which projects are acceptable in the public sector in India is 10 per cent in real terms (Murty *et al.* 1992), this rate has been adopted for the purpose of this study.

4

Cost of the Ganga Action Plan

CAPITAL COST OF THE GAP

The total funds released for investment expenditure of the Ganga Action
Plan (under both the GAP Phases I and II) at 1995–6 prices were Rs
7657.37 million during the period 1985–6 to 1996–7 (see Table 4.1). Of

TABLE 4.1

Investment Expenditure on the Ganga Action Plan

(Rs million)

Year	Current Price			Constant (1995–6) Price		
	Funds released for			Funds released for		
	GAP I	GAP II	Total	GAP I	GAP II	Total
1985–6	65.20	0.00	65.20	153.87	0.00	153.87
1986–7	258.00	0.00	258.00	575.34	0.00	575.34
1987–8	430.00	0.00	430.00	885.80	0.00	885.80
1988–9	568.90	0.00	568.90	1092.29	0.00	1092.29
1989–90	600.00	0.00	600.00	1074.00	0.00	1974.00
1990–1	517.40	0.00	517.40	838.19	0.00	838.19
1991–2	498.10	0.00	498.10	707.30	0.00	707.30
1992–3	536.90	0.00	536.90	692.60	0.00	69.60
1993–4	468.60	179.50	648.10	557.63	213.61	771.24
1994–5	271.29	39.51	310.80	293.00	42.67	335.67
1995–6	123.05	144.17	267.21	123.05	144.17	267.22
1996–7	52.33	211.54	263.87	52.33	211.54	263.87
Total	4389.77	574.72	4964.48	7045.40	611.99	7657.39

the total funds released, Rs 7045.40 million is under the GAP Phase I and the remaining Rs 611.97 million under the GAP Phase II. Therefore GAP Phase I accounts for nearly 90 per cent of the total funds released so far. Funds released under the Ganga Action Plan Phase I by states are given in Table 4.2. This shows that more than 80 per cent of the funds were released for the GAP projects in Uttar Pradesh and West Bengal (each state's share accounts for 40 per cent of total funds), while Bihar accounts for only 12 per cent of the funds released. It is proposed that the remaining 5 per cent excess expenditure is counted under the establishment cost of the GAP Phase II. Total actual expenditure under the GAP Phase I is Rs 6397.25 million (see Table 4.3). This includes Rs 2769.88 million in Uttar Pradesh, Rs 2782.94 million in West Bengal, and Rs 844.43 million in Bihar.

Table 4.3 also provides information about the time phasing of investment expenditures for the GAP Phase I during the period 1985–6 to 1996–7. These expenditures cover a large number of water pollution abatement projects contributing to the clean-up of the Ganges. They have also created employment for a large number of surplus unskilled labourers in the states of Uttar Pradesh, Bihar, and West Bengal in the Gangetic basin.

Estimates based on detailed data on the construction of three Sewage Treatment Plants (STPs) show that the expenditures on skilled and unskilled labour constitute respectively 22 per cent and 24 per cent of the total capital cost of the GAP projects. Out of total capital cost of Rs 7657.37 million at 1995–6 prices, the unskilled labour and skilled labour employed for the construction of the GAP projects amount to Rs 1837.77 million and Rs 1684.62 million respectively. Table 4.4 provides details of the composition of the capital cost of the GAP in terms of domestic material and skilled and unskilled labour.

OPERATION AND MAINTENANCE COST

Expenditures on the operation and maintenance of projects under the Ganga Action Plan Phase I by states can be seen in Table 4.5. During the period 1986–7 to 1996–7, the total operation and maintenance expenditure under the GAP Phase I at 1995–6 prices was Rs 355.70 million. This breaks down to Rs 231.36 million (66 per cent) for Uttar Pradesh, Rs 67.87 million (19 per cent) for West Bengal, and Rs 56.48 million (15 per cent) for Bihar. Table 4.5 also provides the time phasing of operation and maintenance expenditure of the

TABLE 4.2
State-wise Investment Expenditure on the Ganga Action Plan

(Rs million)

Year	Current prices					Constant (1995–6) prices				
	Funds released for GAP I					Funds released for GAP I				
	UP	Bihar	West Bengal	Other GDP etc.	Total	UP	Bihar	West Bengal	Other GDP etc.	Total
1985–6	46.60	4.50	10.00	4.10	65.20	109.98	10.62	23.60	9.68	153.87
1986–7	163.00	24.80	59.20	11.00	258.00	363.49	55.30	132.02	24.53	575.34
1987–8	191.80	63.10	164.30	10.80	430.00	395.11	129.99	338.46	22.25	885.80
1988–9	167.80	124.40	257.90	18.80	568.90	322.19	238.85	495.17	36.10	1092.29
1989–90	222.90	88.00	262.40	26.70	600.00	398.99	157.52	469.70	47.79	1074.00
1990–1	198.50	61.20	243.90	13.80	517.40	321.57	99.14	395.12	22.36	838.19
1991–2	184.50	34.00	259.60	20.00	498.10	261.99	48.28	368.63	28.40	707.30
1992–3	228.90	77.90	214.80	15.30	536.90	295.28	100.49	277.09	19.74	692.60
1993–4	244.00	44.20	154.80	25.60	468.60	290.36	52.60	184.21	30.46	557.63
1994–5	86.50	13.00	147.28	24.22	270.99	93.42	14.04	159.06	26.15	292.67
1995–6	58.16	0.00	40.13	22.45	120.74	58.16	0.00	40.13	22.45	120.74
1996–7	44.15	0.00	0.00	8.18	52.33	44.15	0.00	0.00	8.18	52.33
Total	1836.81	535.10	1814.30	200.95*	4387.16	2954.67	906.83	2883.18	298.09	7042.77

*The excess expenditure is proposed to be booked under the establishment cost of GAP II.

TABLE 4.3
Actual Expenditure on GAP Phase I (as on 01.08.97)

(Rs million)

Year	Current prices				Constant (1995–6) prices			
	UP	Bihar	West Bengal	Total	UP	Bihar	West Bengal	Total
1985–6	8.27	0.00	3.13	11.40	19.52	0.00	7.40	26.91
1986–7	110.39	20.61	60.52	191.52	246.18	45.95	134.96	427.09
1987–8	213.96	75.78	164.22	453.96	440.76	156.12	338.29	935.16
1988–9	170.66	110.55	256.72	537.92	327.66	212.25	492.90	1032.81
1989–90	186.07	75.29	235.96	497.32	333.07	134.77	422.37	890.20
1990–1	175.27	44.54	188.10	407.91	283.93	72.16	304.72	660.81
1991–2	204.59	40.38	254.86	499.83	290.51	57.35	361.90	709.76
1992–3	234.03	58.34	266.20	558.57	301.89	75.27	343.40	720.56
1993–4	224.51	39.69	183.91	448.11	267.17	47.23	218.86	533.25
1994–5	87.72	15.63	95.77	199.12	94.74	16.88	103.44	215.05
1995–6	128.56	20.78	47.29	196.62	128.56	20.78	47.29	196.62
1996–7	35.90	5.71	7.42	49.03	35.90	5.71	7.42	49.03
Total	1780.45	507.29	1764.11	4051.85	2769.88	844.43	2782.94	6397.25

*The excess expenditure is proposed to be booked under the establishment cost of GAP II.

TABLE 4.4

Domestic Material, Skilled Labour, and Unskilled Labour Components of the Capital Cost of the Ganga Action Plan (at 1995–6 prices)

(Rs million)

Year	Domestic material	Skilled labour	Unskilled labour	Total capital cost
1985–6	83.09	33.85	36.93	153.87
1986–7	310.68	126.58	138.08	575.34
1987–8	478.33	194.88	212.59	885.80
1988–9	589.84	240.30	262.15	1092.29
1989–90	579.96	236.28	257.76	1074.00
1990–1	452.62	184.40	201.17	838.19
1991–2	381.94	155.61	169.75	707.30
1992–3	374.01	152.37	166.22	692.60
1993–4	416.47	169.67	185.10	771.24
1994–5	181.26	73.85	80.56	335.66
1995–6	144.30	58.79	64.13	267.21
1996–7	142.49	58.05	63.33	263.87
Total	*4134.99*	*1684.63*	*1837.77*	*7657.37*

TABLE 4.5

Operation and Maintenance Expenditure under GAP Phase I

(Rs million)

Year	Current prices				Constant (1995–6) prices			
	UP	Bihar	West Bengal	Total	UP	Bihar	West Bengal	Total
1986–7	1.139	1.384	0.365	2.888	2.540	3.086	0.814	6.440
1987–8	1.339	4.588	0.458	6.385	2.758	9.451	0.943	13.153
1988–9	6.361	5.102	3.055	14.518	12.213	9.796	5.866	27.875
1989–90	14.569	2.535	3.894	20.998	26.079	4.538	6.970	37.586
1990–1	10.578	1.974	3.082	15.634	17.136	3.198	4.993	25.327
1991–2	26.006	2.223	5.910	34.139	36.929	3.157	8.392	48.477
1992–3	27.733	2.317	7.520	37.570	35.776	2.989	9.701	48.465
1993–4	24.576*	2.765*	8.150*	35.491*	29.245	3.290	9.699	42.234
1994–5	32.469*	4.161*	6.595*	43.225*	35.067	4.494	7.123	46.683
1995–6	26.047	11.601	10.659	48.307*	26.047	11.601	10.659	48.307
1996–7	7.566*	0.875*	2.714*	11.155	7.566	0.875	2.714	11.155
Total	*178.383*	*39.525*	*52.402*	*270.310*	*231.356*	*56.475*	*67.874*	*355.703*

*GDP share as reported by the state yet to be released, except in case of West Bengal.

TABLE 4.6

Capital and O&M Cost of the Ganga Action Plan (at 1995–6 prices)

(Rs million)

Year	Total capital cost	O&M cost under GAP I	O&M cost under GAP II	Total O&M cost GAP I & II
1985–6	153.872	–	–	–
1986–7	575.340	6.440	0	6.440
1987–8	885.800	13.153	0	13.153
1988–9	1092.288	27.875	0	27.875
1989–90	1074.000	37.586	0	37.586
1990–1	838.188	25.327	0	25.327
1991–2	707.302	48.477	0	48.477
1992–3	692.601	48.465	0	48.465
1993–4	771.239	42.234	17.088	59.322
1994–5	335.662	46.683	6.826	53.509
1995–6	267.213	48.307	56.225	104.532
1996–7	263.866	11.155	44.422	55.577
Total	*7657.371*	*355.703*	*124.561*	*480.264*

TABLE 4.7

Domestic Material, Skilled Labour and Unskilled Labour
Components of O&M Cost of the GAP (at 1995–6 prices)

(Rs million)

Year	Domestic material	Skilled labour	Unskilled labour	Total O&M cost
1986–7	5.088	0.258	1.095	6.440
1987–8	10.391	0.526	2.236	13.153
1988–9	22.021	1.115	4.739	27.875
1989–90	29.693	1.503	6.390	37.586
1990–1	20.008	1.013	4.306	25.327
1991–2	38.297	1.939	8.241	48.477
1992–3	38.288	1.939	8.239	48.465
1993–4	46.865	2.373	10.085	59.322
1994–5	42.272	2.140	9.097	53.509
1995–6	82.580	4.181	17.770	104.532
1996–7	43.906	2.223	9.448	55.577
Total	*379.409*	*19.211*	*81.645*	*480.264*

GAP I, at current and 1995–6 prices. Data on the operation and maintenance expenditure of GAP II is not available and, therefore, the estimates of these expenditures are obtained by assuming that the ratio of operation and maintenance cost and capital cost of GAP II is same as that of GAP I as given in Table 4.6. Estimates of the share of skilled and unskilled labour in the total operating cost of the GAP projects based on the detailed data of three STPs and three Municipal Pumping Stations (MPSs) show that they form respectively 4 per cent and 17 per cent of operation and maintenance cost. Out of the total operation and maintenance cost of Rs 480.26 million at 1995–96 prices, skilled and unskilled labour account for Rs 19.21 million and Rs 81.65 million respectively.

COST OF VARIOUS TYPES OF SCHEMES

There are about 261 schemes sanctioned under the Ganga Action Plan so far. These schemes are divided into six types: (a) interception and diversion; (b) sewage treatment plants; (c) low cost sanitation; (d) electric crematoria; (e) river front development; and (f) other schemes. The details of each type of scheme are given in Table 4.8. Totals of schemes in each type are: 88 for interception and diversion, 43 for low cost sanitation, 35 for both sewage treatment plants and river front development, 28 under electric crematoria, and 32 under other schemes. The costs by type of scheme are: Rs 1693.96 million (45 per cent of the total sanctioned cost) for sewage treatment plants, Rs 1448.24 million (38 per cent of cost) for interception and diversion, and Rs 645.19 million (17 per cent of total cost) for low cost sanitation, electric crematoria, river front development, and others.

Among the three states, West Bengal has 110 schemes, Uttar Pradesh has 106, and Bihar has 45. The total sanctioned cost of these schemes is Rs 1787.66 million in West Bengal, Rs 1496.08 million in Uttar Pradesh, and Rs 503.35 million in Bihar. There are a large number of schemes in Uttar Pradesh for interception and diversion and the sanctioned cost of these and other schemes is high when compared with other states. In West Bengal there are a large number of schemes for electric crematoria and river front development, and the sanctioned cost is high when compared with other states. There are also more schemes for sewage treatment plants and low cost sanitation in West Bengal, but the sanctioned cost is low compared to Uttar Pradesh.

TABLE 4.8
Key Facts on Various Types of Schemes

	Uttar Pradesh	Bihar	West Bengal	Total
A. *Intereception and Diversion*				
1. Number of schemes sanctioned	40	17	31	88
2. Sanctioned cost (Rs million)	416.54	196.99	834.70	1448.24
3. Number of pumping stations	33	27	69	129
(a) Number completed	25	16[+]	8	49
(b) Yet to be completed	8	11	61	80
4. Length of sewer (km) (including force main)	136 (45)	55 (7)	179 (26)	370 (78)
(a) Quantity laid	136 (45)	54 (7)	170 (23)	360 (75)
(b) Quantity yet to be laid	0 (0)	1 (–)	9 (3)	10 (3)
5. Total wastewater from Class I towns (mld)	680	133	527	1340
6. Quantity of wastewater to be intercepted/diverted (mld)	431	115	327	873
7. Capacity commissioned for interception (mld)	401	99.2	211.75	711.95
8. Capacity commissioned for treatment (mld)	185.3	99.2	200	484.5
9. Schemes completed	40	17	28	85
10. Schemes ongoing	–	–	3	3
B. *Sewage Treatment Plants*				
1. Number of schemes sanctioned	13	7	15	35
*Renovation to capacity	–	–	2	2
*Renovation and Expansion	2	2	4	8
*New	11	5	9	25
2. Sanctioned cost (Rs million)	767.71	190.16	736.09	1693.96
3. Total STP capacity (mld)	375	141.5	372	888.5
4. Schemes completed	10	3	12	25
5. Schemes ongoing	3	4	3	10
C. *Low Cost sanitation*				
1. Number of schemes sanctioned	14	7	22	43
2. Sanctioned cost (Rs million)	100.07	55.03	67.92	223.21

	Uttar Pradesh	Bihar	West Bengal	Total
3. Number of individual toilets				
(a) Sanctioned	24,965	6725	20,710	52,400
(b) Completed	(20,282)	6725	20,698	47,705
4. Number of community toilets				
(a) Sanctioned	189	116	2458	2763
	(2440)	(1528)	(2724)	(6692)
(b) Completed	183	116	2458	2757
	(2360)	(1523)	(2724)	(6607)
5. Schemes completed	14	7	22	43
6. Schemes ongoing	–	–	–	–
D. *Electric Crematoria*				
1. Number of schemes sanctioned	3	8	17	28
2. Sanctioned cost (Rs million)	19.68	40.42	78.15	138.24
3. (a) Number of crematoria sanctioned	4	9	20	33
(b) Number of crematoria completed	4	9	17	33
(c) Number of crematoria under construction	–	–	3	3
4. Schemes completed	3	8	15	26
5. Schemes ongoing	–	–	2	2
E. *River Front Development*				
1. Number of schemes sanctioned	8	3	24	35
2. Sanctioned cost (Rs million)	63.76	8.75	67.76	140.27
3. Number of ghats to be developed	44	10	75	129
4. Number of ghats developed	43	10	75	128
5. Schemes completed	8	3	24	35
6. Schemes ongoing	–	–	–	–
F. *Other Schemes*				
1. Number of schemes sanctioned	28	3	1	32
2. Sanctioned cost (Rs million)	128.33	11.98	3.05	143.36
3. Schemes completed	28	3	1	32
4. Schemes ongoing	–	–	–	–

Notes: + In addition, 4 PSs have been completed except pumps' installation.

* In addition, 2 STPs in Bihar (Beur and Saidpur) and 2 STPs in West Bengal (Bhatpara and Titagarh) have been renovated and recommissioned. Further augmentation of these plants is in progress.

Figures in parentheses under Section A represent the force main.
Figures in parentheses under Section C represent the number of seats.
mld: million litres per day.

EFFLUENT TREATMENT COST OF GROSSLY
POLLUTING INDUSTRIAL UNITS IN GANGETIC BASIN

There are about 68 heavily polluting industrial units in the Gangetic basin, out of which 26 are central public sector units, 6 are state public sector units, and the remaining 36 are private sector units. These 68 units discharge around 260 million litres of effluent every day. Table 4.9 provides information about the comparative status of installation of effluent treatment plants (ETPs) in the industrial units. During the post-GAP period the total number of industrial units having ETPs has gone up from 14 to 55.

TABLE 4.9

Comparative Status of Installation of ETPs by Grossly
Polluting Industries in the Ganga Basin

State	Unit with ETPs		ETPs under construction		Unit without ETPs/ not required		Unit closed	
	Sept. 1985	June 1995	Sept. 1985	June 1995	Sept. 1985	June 1995	Sept. 1985	June 1995
UP	2	28	15	0	15	0	2	6
Bihar	2	4	2	0	1	0	0	1
West Bengal	10	23	5	0	11	1	3	5
Total	14	55	22	0	27	1	5	12
CU	4	22	8	0	13	0	1	4
SU	0	6	2	0	4	0	0	0
PU	10	27	12	0	10	1	4	8

CU = Central Public Sector Units; SU = State Public Sector Units; PU = Private Sector Units.

Data was collected for the annual turnover, capital, and operating cost of ETPs, effluent volume, and effluent characteristics for a sample of 18 water polluting industries in the Gangetic basin. Table 4.10 provides this information. The effluent treatment cost per kilolitre of effluent is estimated as Rs 0.39 at 1993–4 prices, using the data for industrial units in the sample. Given that the volume of effluent discharge from 68 heavily polluting industrial units is 2.6 million kilolitres per day, the daily cost of treating the effluents is estimated as Rs 1.014 million at 1993–4 prices. Using these estimates, the annual cost of

TABLE 4.10

Information of Turnover, Estimates of ETP Cost, Wastewater Volume, and Influent and Effluent Characteristics of a Sample of Water Polluting Industries in the Gangetic Basin

Factory code	Turnover (Rs mn)	Total ETP cost (Rs mn)	Volume of untreated water (m^3/day)	Waste-water capacity (m^3/day)	Influent			Effluent		
					BOD	COD	SS	BOD	COD	SS
Fertilizer										
1	289.90	9.37	40	40	30,000	100,000	200	100	250	80
2	2795.53	17.57	9960	11550	50	–	–	15	–	–
3	131.08	3.43	8400	–	22	165	50	23	140	28
Sugar										
4	57.74	0.30	400	500	–	–	–	–	–	–
Distillery										
5	211.59	13.06	240	250	24,800	66,000	3280	22	84	37
6	404.90	404.90	160	300	48,000	78,250	2470	35	225	65
Chemical										
7	242.27	0.97	116	240	10	20	76	13	55	50
8	2200.00	9.47	1860	1860	1000	4810	225	27	200	75
9	1479.70	5.22	16,000	175,000	45	150	160	18	90	40
10	4500.00	14.67	250	360	80	1568	184	28	149	46

Factory code	Turnover (Rs mn)	Total ETP cost (Rs mn)	Volume of untreated water (m³/day)	Waste-water capacity (m³/day)	Influent			Effluent		
					BOD	COD	SS	BOD	COD	SS
Refinery										
11	9280.10	20.00	6000	12,960	60	–	50	12	114	17
12	5076.35	4.07	10,000	14,400	115	275	80	5	30	15
Tannery										
13	263.85	6039	800	1600	157	307	147	22	205	78
Paper										
14	80.59	2.23	1000	2000	150	450	200	26	140	40
15	1400.00	18.66	18,000	20,400	361	1108	530	25	540	80
Miscellaneous										
16	3880.00	3.43	25,000	28,000	53	128	105	14	28	35
17	1149.52	0.24	76	87	372	450	140	20	90	50
18	557.80	5.10	450	150	25,000	40,000	400	25	240	–

Note: National Standards for BOD = 30 mg/l, COD = 250 mg/l, and SS = 100 mg/l.
Source: Survey data.

effluent treatment for heavily polluting industrial units is estimated as Rs 370.11 million at 1993-4 prices.

CONCLUSION

This chapter has reported the cost details of the GAP in terms of the actual costs incurred and those costs at constant prices, that is adjusting for inflation. Both these figures are referred to as financial costs. In addition, one can look at costs in economic and social terms, that is in terms of social values of the expenditures because market prices do not always measure social costs. The economic costs corresponding to these financial costs are given in Chapter 12. The differences are not great; in general the economic costs are about 6–16 per cent higher for investment costs and about 24 per cent higher for operating costs. Both financial and economic cost data are used in the analysis of net benefits in Chapter 12.

5

Measuring Non-user Benefits from Cleaning-up the Ganges

INTRODUCTION

This chapter evaluates the benefits of the GAP to those who do not normally reside near the river, and who do not use the river directly. Such benefits are otherwise called non-user benefits or existence and bequest values as described in Chapter 3. There can be several motives why people who may not be planning to visit the Ganges in their lifetime express satisfaction from the knowledge that the Ganges is clean.

First, one may feel a general sense of satisfaction from just knowing that a major river in the country is now cleaner than it was earlier, whether this stems from ecological considerations or from the point of view of pure scenic beauty. Second, there can be a sense of satisfaction from knowing that fellow citizens living near the river have a better quality of life with a cleaner river. Third, there could be some perceived benefits from knowing that a river of major religious significance is now cleaner. Finally, there is an element of consideration for future generations, in that there is a sense of enhanced well-being from knowing that future generations will inherit a cleaner river resource in the future.

This chapter begins by describing briefly the use of the benefit estimation technique known as the contingent valuation method, which is now employed widely to measure environmental benefits, before going on to detail the specific methodology used in the contingent valuation survey of non-user benefits from the GAP. The more technical aspects of the methodology used are presented in Appendix 5.1. Details of the calculation of the River Water Quality Index used in the estimations are provided in Appendix 5.2.

VALUATION OF ENVIRONMENTAL BENEFITS USING THE CONTINGENT VALUATION METHOD

Contingent valuation surveys simulate a market for a non-marketed good providing environmental services, and thus obtain a value for the good contingent on the hypothetical market described during the survey. The design of the questionnaire to elicit individual responses to the valuation question is thus of central importance in a contingent valuation survey. Not only must the information elicited be accurate, but it must be relevant to the valuation problem at hand. All contingent valuation studies yield data and information. However, a poorly designed study can lead to information that suffers from several biases, including strategic biases—which mean respondents have tried to conceal their true preferences (and thus have given dishonest values)—or biases resulting from the good in question or the hypothetical market being improperly described to the respondent. There may also be biases resulting from improper sampling design or execution, and those from improper aggregation.[1] Since there is no 'true' value of the good in question (as determined, for instance, by a market), it is difficult to distinguish a value presented in a poorly-designed study from that coming out of a well-designed one, except by examining the details of the study—including the questionnaire used, the sample covered, and methods used to aggregate individual responses to represent the population sampled.

A contingent valuation survey consists of several components:

(1) a *questionnaire*, which includes a clear description of the public good (or environmental amenity) to be valued, the method of eliciting values, and questions on the socio-economic status of the respondent;

(2) *sampling* from the target population, by suitably-trained enumerators who interview a sample chosen to be as representative of the population as possible; and

---

[1]Mitchell and Carson (1989) identify four principal sources of bias:

(1) use of a scenario that contains strong incentives for respondents to misrepresent their true WTP amounts; (2) use of a scenario that contains strong incentives for respondents to improperly rely on elements of the scenario to help determine their WTP amounts; (3) misspecification of the scenario by incorrectly describing some aspect of it, or, alternatively, by presenting a correct description in such a way that respondents misperceive it; and (4) improper sampling design or execution, and improper benefit aggregation. [p. 235].

See their Table (11.1) on pp. 236–7 for a more complete listing of possible biases in contingent valuation studies.

(3) *data analysis*, which includes deriving average (or household) willingness-to-pay (WTP) estimates from the sample information, and then extrapolating household estimates to the entire population.

Each of the major components of a contingent valuation survey are discussed in greater detail below in the specific context of the valuation of non-user benefits from the Ganga Action Plan.

CONTINGENT VALUATION OF NON-USER BENEFITS OF THE GAP

Following an outline of the procedure followed in conducting the contingent valuation survey of non-user benefits from the Ganga Action Plan, detailed description of the questionnaire and the sample covered is presented below.

SALIENT FEATURES OF THE CONTINGENT VALUATION SURVEY

Since one of the aims of the exercise was to provide an estimate of the benefits from the improvement in river water quality from 1985 to 1995, a part of the benefit estimation had necessarily to be *post hoc*, meaning that respondents had to be asked their willingness-to-pay for benefits which they had *already received*. This represents a departure from the conventional contingent valuation practice. This problem was tackled by asking respondents to simply value the benefits they felt they would have received from two levels of water quality (shown in maps), one as it was in 1985, and the other showing the situation in 1995.

Because the idea of valuing a river that was 'half-clean' or 'three-quarters' clean is conceptually difficult for the respondents, it was decided first to ask them their willingness-to-pay for best water quality, which was defined as 'uniform bathing quality throughout the river', based on a map showing uniform bathing quality. The idea was to get the respondents to provide a benchmark estimate for *best* water quality, which they could use subsequently to provide values for water quality as it was in 1995 and in 1985. Maps of river water quality were drawn using data from the National Rivers Conservation Directorate (see Figures 1.2, 1.3, and 1.4), and respondents were asked to make their evaluations of three water quality levels using three maps: first for best quality, then for the 1995 quality, and finally for the 1985 quality. Given separate values for each scenario, several incremental values can be inferred subsequently. For example, subtracting the value given for the

1985 quality from that given for the 1995 quality would give the incremental value of the benefits of the improvement in water quality from 1985 to 1995, one of the target questions of the overall project.

Lastly, the elicitation format (detailed in Table 5.1) used a variant of the open-ended bidding game format, backed by a version of a 'Payment Card', in a multiple bid format with an initial bid, a second bid, and a final bid. The first bid was open-ended, but based on a 'Payment Card' (to avoid the problem of a high percentage of protest bids, faced by early contingent valuation surveys; Mitchell and Carson, 1989). But since this could result in a range bias, a second take-it-or-leave-it question, based on the initial bid was asked. Possible initial bids were classified into five class intervals, and the amount used in the follow-up question was twice the mid-point of the class interval. A final question asked for the maximum they would pay, and the response to this question was taken to be the WTP value.

Care was taken for the first value elicited (for benefits from bathing quality water in the river) since this was to be the benchmark valuation. Once the three questions leading to the maximum WTP for benefits from bathing quality were asked, respondents were given the second map of water quality (1995 level) and asked simply for their maximum WTP for the benefits they perceived they would receive (for the entire household per annum) as non-users. Respondents were then given a map of river water quality as it existed in 1985 and were asked their maximum WTP for perceived benefits from this third level of river water quality.

THE QUESTIONNAIRE

The questionnaire was developed by researchers of the Institute of Economic Growth, in collaboration with Metroeconomica of the United Kingdom. Following an initial workshop to discuss methodology, the questionnaire was modified in response to comments from international experts on the contingent valuation method, and pre-testing was carried out in six cities in the country. The final questionnaire had the following features.

Information about GAP

This first section carried three basic kinds of information about the GAP thought to be essential to the knowledge base required to answer the valuation questions: first, an overview of the GAP, including the year of its inception and target (uniform bathing quality in the river); second, a brief summary of the basic causes for pollution in the river, pointing out that domestic sewage from large cities was the single largest cause for pollution (and not industrial pollution, as is commonly understood); and third, a short account of the activities of the GAP, ranging from the setting up of sewage treatment plants (to combat sewage-based pollution) to constructing electric crematoria (to control the deposits of unburnt or partly-burnt corpses in the river).

Short, multiple-choice questions followed, designed to deliver additional information in small doses which are easier to assimilate. These ranged from whether or not the respondent had visited the river previously and observed its water quality, to short questions about which of the various non-user benefits associated with a cleaner river the respondents identified with. The latter were carefully worded so as to make the respondent think about each non-user benefit, which in turn would be crucial in the subsequent questions on valuation.

Preference Elicitation

This section carried questions designed to make the respondent think about the responsibility of action to clean the river. In other words, it aimed to elicit consumer preferences for action to clean the river. The concluding question, designed to lead on to the valuation section, asked respondents what role they perceived for themselves as individuals in the fight against pollution, irrespective of governmental efforts in that direction.

Value Elicitation

This was the most crucial section of the questionnaire and carried questions asking respondents how much they would be willing to pay for different levels of water quality.[2] The various components of this section are described in detail below.

(a) POSING THE VALUATION QUESTION: The framing of the value elicitation questions for the valuation of benefits of improving river water quality from 1985 to 1995 posed a credibility problem. Since the GAP and its funding was public knowledge ten years after its inception, it was difficult to find an appropriate and credible scenario for eliciting

---

[2]The choice of asking willingness-to-pay questions rather than willingness-to-accept was put forward in the initial design of the questionnaire by the questionnaire preparation team at the Institute of Economic Growth, and ratified by both Metroeconomica and the international expert whose comments were solicited.

payment values from citizens for benefits they had *already* received (that is for water quality improvements between 1985 and 1995). There was no really satisfactory answer to the natural question of 'willing to pay for what?'. For, if the justification was to pay for improvements in water quality, respondents could point out that since the improvements had *already* taken place, where was the necessity for further payments? If the stated justification, say, was to repay loans taken for the work undertaken in these ten years, there was always the danger of someone pointing out that most of the funds were internally generated, and that being government allocations there were no interest payments or repayments outstanding.

It was therefore decided to pose the question as an explicit valuation problem. Respondents were shown a certain map of water quality (either of uniform bathing quality, or river water quality as it was in 1995 or in 1985) and explicitly asked to assume (1) that the water quality shown was actually obtained at the present, and (2) that they should expect no further changes in it. They were then asked to evaluate the benefits they perceived themselves (and their households) as receiving currently, and would receive, as non-users if water in the river stayed at that particular quality level. In other words, posing the question in this format meant that there was a credible reason for soliciting payment: it was a payment to enjoy the benefits that respondents perceived themselves as receiving in the hypothetical situation that the river water quality was at a particular level. Since the various non-user benefits they could expect to enjoy had already been discussed in detail in the first section of the questionnaire, reference back to the answers given in that section refreshed their perception of these benefits. And given the introductory statements in the Value Elicitation Section, on the relationship between willingness to pay for benefits received, and the explanatory statements on budget constraints, this seemed a plausible format.

Also, the fact that respondents were explicitly asked to assume that the water quality shown in the map actually existed, and that they could expect no further changes from it, helped to avoid the tendency among some respondents, which was detected during pre-testing, to value worse water quality higher—since they imagined they would have to pay more to clean up a dirtier river. The revised statement thus stated: 'Assume this [map of either water quality in 1985, 1995, or ideal quality] is the quality of the water in the river right now, and that there will be no further change in the quality . . . . ' This enabled respondents to focus on the problem of valuing their non-user benefits from a river of such water quality, instead of being diverted to the problem of paying to clean up the river.

(b) DETAILING THE HYPOTHETICAL SCENARIO: The major difficulty here was to provide a plausible scenario to the respondents, that is describing the hypothetical market within which the valuation was to take place. The problem was twofold: (1) to acquaint respondents with the idea of stating intentions to pay for indirect benefits from certain levels of water quality; and (2) to get respondents to provide a money value for the same.

In order to tackle the first part of the problem, respondents were given examples of private goods to illustrate the link between willingness-to-pay for benefits received (for example we are willing to pay Rs X for a pen because we expect at least Rs X worth of benefits in return). Once this was grasped, the 'missing market' characteristic of the public good of water quality was explained, which made its valuation possible only by directly asking respondents what value they would place on the benefits they received (as non-users). Being convinced of the validity of providing subjective estimates of value for non-user benefits from improved water quality was not, however, sufficient for respondents to come up with an actual money value for these benefits.

The second part of the problem was addressed by asking respondents to consider river water quality as a sort of public good, in the sense that it was provided by the government for use by all citizens. And for the various other public goods that the government also provided, such as educational facilities, electricity and power, health facilities, and transport, a rough measure of the benefits received per household was the amount the government spent per year per household to provide these various public goods. A card was then shown to the respondents, detailing the expenditure made by the government in the 1990s per household to provide various public goods, to facilitate the estimation of non-user benefits from the given map of river water quality.

(c) SPECIFYING THE BUDGET CONSTRAINT: Two other explanatory statements were made. First, that any payment would have to take income and other spending constraints into account. Second, that while one could consider the benefits of good health as 'priceless' or 'infinite', in practice, one could not spend more than one's income on health facilities, and in reality, one only spent a small proportion of one's income on such facilities; and that this was the sense in which the respondents were being asked to evaluate the benefits they would receive from improved water quality. Together with the reminder of the existence of an income con-straint, this latter clarification was designed expressly to avoid impossib-ly high estimates that may be expected in the context of a river with considerable religious and symbolic importance.

In addition, all respondents who answered 'no' to the Respondent Evaluation Question: 'Do you feel you may actually be required to pay for water quality improvements in the Ganga?', were dropped from the sample. The reason for this was that those who did not feel that their answers had to be constrained by considerations of actual payment may have deliberately overstated their willingness-to-pay.

(d) THE CHOICE OF THE PAYMENT VEHICLE: Of the three payment vehicles considered, payment to the government was rejected outright at the pre-testing stage, given the prevailing air of a lack of confidence in the government's ability to channel the funds in the right direction without 'leakages'. The other two payment vehicles tried in the pre-test-ing were (i) payment to a reputable charitable organization, and (ii) pay-ment to a local citizens' group (of which the respondent could become a member). A reputable charitable organization was the most acceptable payment vehicle and was therefore chosen.

(e) THE CHOICE OF ELICITATION FORMAT: As stated earlier, an open-ended bidding game was used with a variant of the payment card, and respondents were asked two follow-up questions to their initial (non-zero) bid, one of which was a close-ended question. The general features of the elicitation format are described in greater detail below, while the nature of this particular elicitation format is presented in Table 5.1.

An open-ended bidding game is where respondents are initially asked 'How much would you pay [for the good in question]?', whereas in a close-ended bidding game, respondents are initially asked 'Would you pay Rs X [for the good in question]?'. The initial questions in an open-ended bidding format can be followed up (in the case of a non-zero initial bid) with (further questions leading up to) the final question: 'What is the maximum you would pay [for the good in question]?'. In the case of close-ended bidding games, the initial question can be fol-lowed up with further questions based on the answer to the initial ques-tion. If the answer to the initial question was a 'Yes', then the follow-up question could ask: 'Would you pay Rs X + an increment [for the good in question]?'; but if the answer to the initial question was a 'No', then the follow up question could be: 'Would you pay Rs X – some fixed amount [for the good in question]?'

Although there is currently a debate on whether open-ended or close-ended bidding should be used, there appears to be insufficient cause to reject the former in favour of the latter.[3] But while the use of the open-ended bidding game seems justified, there is a longer-standing problem with open-ended questions, which is the high proportion of 'zero bids' they evoke, partly because of the difficulty respondents face in pulling a number out of their heads. This problem has however been addressed successfully by Mitchell and Carson (1981, 1984), by using a 'Payment Card'. A 'Payment Card' carries values of comparable public goods (say, paid for by tax revenues) intended to provide respondents with a rough idea of the range of comparable benefit estimates. The use of a 'Payment Card' helps focus respondents and elicit a much higher proportion of valid (non-zero) initial responses. But the open-ended bidding game backed by a 'Payment Card' has been criticized on the grounds that the resulting values tend to be limited by the range presented in the card.

The elicitation format used in this survey began by describing the good in question and the logic of eliciting willingness-to-pay estimates from the general public (described above). Respondents were then shown a map of the river Ganga with uniform bathing quality water and asked: 'How much would you pay per year for the non-user benefits

[3]The oldest and most widely used format, the open-ended bidding game (where respondents are simply asked to state a value) has been criticized on one major count: a 'starting point bias' can result if starting points are given (e.g. respondents are simply asked whether or not they are prepared to pay a particular price for the good in question). Mitchell and Carson (1981, 1984) suggested tackling this problem by using a 'payment card' with a range of certain comparable values, but in the use of a set of values on a payment card, the mere presence of these values on a card can lead to a 'range bias', where all values given by the respondents are contained within the range of numbers given on the card. It was to try and overcome this problem that the present study used a payment card to get initial values and then asked two follow-up questions to help respondents reach a final value. Another point of debate is the use of close-ended versus open-ended bidding formats, with the Blue Ribbon Panel of the United States National Oceanic and Atmospheric Administration (NOAA) recommending the use of close-ended bidding games as the elicitation format, on the grounds that 'there is no strategic reason for the respondent to do otherwise than answer truthfully' (Arrow *et al.*, 1992: p. 21). However, both Fisher (1996: pp. 24–6) and Diamond (1996: p. 63) argue that 'the case for close-ended CV responses being free of strategic bias has not been made either in theory or by empirical findings'. Thus, while not defending open-ended bidding formats, they merely illustrate that the alternative is not free from the same defect. The choice between the two is therefore not as clear-cut as once assumed.

you (and your family) would derive from a river with this level of water quality?'. The present format also sought to correct for possible range bias by adding two follow-up questions to the initial open-ended question. The first asked a closed-ended question *based on the answer to the initial open-ended question*. The questionnaire had a table detailing the amount to be asked in the first follow-up question, based on the amount offered as the initial bid. This was followed by a second open-ended follow-up question that asked the maximum amount the respondent was willing to pay for the non-user benefits of a river Ganga which had been cleaned up to bathing quality throughout. The bidding format is specified in greater detail in Table 5.1.

For the other two levels of water quality (1995 and 1985), only the final open-ended question wanting to know the maximum willingness-to-pay was asked.

TABLE 5.1

The Elicitation Format: Structure of the Bidding Game

Initial question (open-ended with a variant of the payment card)	*Initial (non-zero) response range*	*Follow-up questions*	
		Question 1 (close-ended but based on the initial response)	*Question 2 (open-ended)*
How much money would you and your family pay per year to enjoy the non-user benefits of a river Ganga which is of bathing quality throughout?		Since you are probably doing such a valuation exercise for the first time, let me ask you: Would you pay Rs	
	Rs 0–250	300	
	Rs 250–500	750	
	Rs 500–1000	1500	
	Rs 1000–2000	3000	
	Above Rs 2000	5000	
		per year on behalf of yourself and your family in order to enjoy the benefits of a river Ganga that is of bathing quality throughout?	
			What is the maximum you would pay for these benefits?

Socio-economic Details

The fourth section of the questionnaire was designed to collect information on socio-economic variables to be used in the regression estimation of the valuation function. Apart from the name of the respondent and the address, information was collected on age, educational level, occupation, size of the household (both above and below the age of 18 years), and gross annual household income from all sources. The last was posed alternatively in the form of bands of possible incomes, which the respondent could choose. Given the proclivity of underestimating income in the country, the subsequent analysis took the upper limit of the income class chosen as representative of the income of the respondent.

Respondent's Evaluation

Several questions were asked in this section as a follow-up to the questions posed in the previous sections. In particular, they asked for the respondent's evaluation of the clarity of the enumerator's presentation of the questionnaire, the extent to which the respondent believed the answers provided would influence policy and affect clean-up operations, and whether or not the respondent believed actual payments would be asked for. These helped subsequently to screen out questionnaires containing responses which may be biased for a lack of clarity of exposition by the enumerator.

Enumerator's Evaluation and Declaration

Two questions were put to the enumerators at the end of the interview, to ascertain the attitude and capability of the respondents to answer the elicitation questions. The enumerators were asked to describe the kind of effort the respondents made to focus on the hypothetical scenario and to understand the context and purpose of the elicitation questions. Those questionnaires where enumerators felt the respondent either did not pay sufficient attention to the valuation questions or did not understand the questions were eliminated from the sample subsequently.

Further, enumerators were asked to sign a declaration stating that the interview had been conducted honestly and in accordance with the instructions they had received during their training. The basic purpose of these questions was to instil a sense of responsibility in the enumerators and to make them exert their utmost to elicit accurate responses. Since the questions probe subjective evaluations by the respondents, there are no right or wrong answers to any of the questions, and experienced but dishonest enumerators could have filled in each questionnaire on their own without actually having carried out the interviews. The only way to check this is by close supervision. The questions in this section of the survey and the declaration accompanying them attempt to check such behaviour on the part of the enumerators. The selection of enumerators is another vital aid to averting such behaviour. This is further explained below.

THE SAMPLING

Being a nationwide survey, the choice of the sample was an important one. There were, however, constraints on the extent of coverage of the whole population posed by the available resources of time and funding. In addition, there was the problem of getting meaningful responses from some groups in the country. For two major reasons, therefore, the sample was restricted to urban populations in the major cities in the country. First, contingent valuation involves considerable hypothetical reasoning, which, it was felt, would be difficult for illiterate respondents to follow and use. Second, willingness-to-pay is constrained by income, and if incomes are low or at subsistence levels, it may be difficult for respondents to truthfully reveal positive values. For these reasons, it was decided to restrict the target population to the literate and employed population in major cities (population of 1 million and above) in the country.

Cities were selected to achieve maximum geographical coverage. Although it would have been ideal to cover all the state capitals in the country, resource constraints meant that outside the major metropolitan centres coverage had to be restricted to simply ensuring that the four geographical zones were adequately covered. For this reason, the cities of Trivandrum (Thiruvananthapuram), Bangalore, and Hyderabad in the south, Baroda in the west, and Allahabad, Lucknow, and Kanpur in the north were chosen, apart from the metropolitan cities of Delhi, Calcutta, and Chennai. Initially, 250 questionnaires per city were planned, totalling 2000, but information from only 1876 filled-in questionnaires was entered into the spreadsheet, the rest being either protest bids or incompletely filled-in questionnaires.

Within each city, the sampling scheme was designed to control for geographical spread and income categories. Accordingly, the city was divided into four geographical zones, and households in three income categories of high-income, middle-income, and low-income groups were sampled. The sampling in each geographical zone and income group was done by taking every alternate household along a street or in a complex of residential flats. Although this is not a stratified random sample as defined in textbooks, given the size of the target population and the enormous costs of information and processing involved in working out population strata, and thence, sample strata (especially if the strata are as diverse as age, sex, occupational groups, social classes or castes, and religion), the methodology adopted appeared to be a cost-effective means of getting a fairly representative sample.[4] Also, for reasons mentioned above, only the literate and the employed were chosen from among the urban population.

The choice of enumerators was restricted to college and university students for a variety of reasons. Apart from the fact that such a survey would give young students (of Economics and Statistics) valuable practical experience with conducting surveys, and that it provided an introduction to applied environmental economics to interested students, it also avoided problems of dishonest reporting of the sort mentioned above.[5] Each enumerator was paid Rs 25 per questionnaire, fully filled-in, plus Rs 10 per questionnaire for local travel costs.

The following section presents the analysis of the data and the results.

ESTIMATION RESULTS

There are three sets of results. First, the calculation of the mean willingness-to-pay (WTP). Second, the extrapolation of these average WTP

[4] It would have been preferable to use electoral rolls to identify the target population and make the sample selection completely random. This, unfortunately, could not be done, given the time constraints.

[5] It is interesting to note that Mitchell and Carson (1995) recommend the use of professional interviewers. They note:

We advocate the use of professional interviewers because they are instructed to adhere strictly to the text of the instrument, whereas graduate students or other types of interviewers who might be recruited on a one time basis are easily tempted to adapt the wording to the respondent or to explain the meaning of text which puzzles the respondent. This type of intervention destroys the purpose of a survey which is to obtain information from a sample of people in response to material which is presented consistently. [p. 28]

They note, however, in their earlier work (1989) that while the field investigating manuals state that 'ad-hoc explanations of what a question means, however well-intended, destroy comparability . . . . *Studies have shown that these goals are not consistently met, even with well-trained interviewers*' (p. 240; emphasis added).

estimates to the population. Third, the calculation of incremental WTP for different sets of changes in water quality.

MEAN WILLINGNESS-TO-PAY

Data from the city samples (Delhi, Chennai, Calcutta, Thiruvananthapuram, Baroda, Bangalore, Hyderabad, Vijayawada, Kanpur, Lucknow, and Allahabad) were pooled to get an 'All Non-users' sample. This data was screened to eliminate incomplete questionnaires including protest (zero) bids and those which did not reflect the budget constraint. It was also screened to eliminate questionnaires where respondents were assessed by the enumerators as either not understanding the valuation question or not giving it sufficient consideration. This reduced the final sample size from 1876 to 817. While this means that roughly 50 per cent of the sample had to be dropped, this is not unusual in Contingent Valuation Surveys, although the average figure is much lower (usually around 10–20 per cent). But the decision to screen the data thoroughly was taken with a view to presenting as accurate an estimate of non-user benefits as possible. In any case, a total sample of over 800 observations is usually deemed adequate for a contingent valuation survey.[6]

Household values for the benefits from three different levels of water quality (best or bathing quality, 1995 quality and 1985 quality) are calculated in four different ways:

(1) as the sample mean;

(2) as the sample median;

(3) by estimating a regression equation with WTP for each level of water quality regressed on a vector of socio-economic variables and dummy variable, and using the average values of the regressors to generate estimates of the dependent variables; and

---

[6]The largest sample size reported in a list of selected CV studies in Mitchell and Carson (1989: Appendix C, Table C-1, p. 363) is 748. See Mitchell and Carson (1989), chapter 12, for a thorough discussion of 'non-sample response' and 'sample selection' biases. Their Table 12-3 on page 281 records response rates of between 8 and 93 per cent, but these have not been corrected for questionnaires where the responses to the CV elicitation questions were missing. The final percentages, thus, would be lower. But note that in the case of the present survey, the sample of useable questionnaires was also reduced because of internal checks to ensure that the budget constraint did actually constrain the responses to the willingness-to-pay questions. The sample size also corresponds roughly to a situation where (a) 95 per cent of the time the estimated mean WTP is within 20 per cent of the true WTP, and (b) the coefficient of variation is about 3.0 (Mitchell and Carson, 1989: pp. 224–5).

(4) by estimating a regression equation with pooled data on WTP for all levels of water quality regressed on an index of water quality, a vector of socio-economic variables and dummy variables, and using the average values of the regressors and the actual value of the water quality index to generate estimates of the dependent variable.

The sample mean, being nearly identical to the estimated mean, is not reported here. The sample median is usually used to provide a measure of central tendency for the sample, when the sample is believed to be highly skewed, since it less influenced by outliers.

Hence, three estimates of household WTP for each of the three water quality levels, namely, (1) BESTMAX for Best or Bathing Quality, (2) CURRMAX for Current or 1995 Water Quality, and (3) PASTMAX for Past or 1985 Water Quality, are given.

ECONOMETRIC MODEL WITHOUT WATER QUALITY

In the regression equation estimated for each level of water quality, the dependent variable is the WTP elicited from the respondents for that particular level of water quality. The vector of socio-economic variables on which these are then regressed includes income, age, education level, and household size, while dummies were added for specific responses given in the questionnaire and for cities where the respondents live (described in detail below).

Three functional forms were tried for each equation—the Cobb–Douglas form (in logarithmic-linear form), a modified Cobb–Douglas form (where income and age entered in squared form), and the Translog (flexible) functional form.[7] The final equation used in the calculation of the mean value was chosen on the basis of the highest Adjusted R^2 and F-statistics (see footnote 12 for an explanation of these statistics). The full specification of each equation in general and in the Cobb–Douglas form follows.

---

[7] The Translog form does not seem to have been used in the specification of WTP functions in previous CV studies. The chief reason for using the Translog specification in the present study is to see if non-linearities in the functional relationship could be captured. There are, however, standard cautions to be applied to these estimates. The main caveat is that these results hold only locally, i.e. within a small neighbourhood of the point of tangency between the possibly non-linear function and the tangent to the function at the point of evaluation. The Translog specification is, as is well known, the linear form of a second-order Taylor expansion at the point of tangency.

General Form

WTP = f(CONSTANT, INCOME, AGE, SIZE, EDU, INF_CL, VISIT,
    B_STAND, D_STAND, CHENNAI, DELHI, TVM, BLE,
    BARODA, KLA, HYV),   (5.1)

where WTP is BESTMAX for best or bathing water quality;
    CURRMAX for current or 1995 quality (with GAP); and
    PASTMAX for past or 1985 quality.

Cobb–Douglas Logarithmic Linear Form

BLOG (or CLOG or PLOG) = $\alpha + \alpha_I$ ILOG + α_A ALOG + α_S SLOG
    $+ \alpha_E$ ELOG + β_1 VISIT + β_2 INF_CL
    $+ \beta_3$ B_STAND + β_4 D_STAND
    $+ \delta_1$ CHENNAI + δ_2 DELHI + δ_3 TVM
    $+ \delta_4$ BLE+ δ_5 BARODA + δ_6 KLA
    $+ \delta_7$HYV + ε,   (5.2)

where

BLOG (or Log BESTMAX) = Log (WTP for Best Quality);
CLOG (or Log CURRMAX) = Log (WTP for 1995 Quality);
PLOG (or Log PASTMAX) = Log (WTP for 1985 Quality);
ILOG (or Log INCOME)   = Log (per capita annual household
                income from all sources);
ALOG (or Log AGE)   = Log (Age of the respondent);
SLOG (or Log SIZE)   = Log (Size of the household: with
                members below the age of 18
                converted to adult equivalent units);
ELOG (or Log EDU)   = Log (Education level of the respond-
                ent, in years)[8]
VISIT   = Dummy variable: Whether respond-
                ents had visited the Ganga in the
                past 10 years: YES = 1; NO = 0;
INF_CL   = Dummy variable: Whether or not
                respondents had heard about the
                Ganga Action Plan; YES = 1; NO = 0;

---

[8]For the translog functional form, the following additional variables were used:
    ILOGI = (ILOG × ILOG)/2
    ILOGA = ILOG × ALOG
    ILOGE = ILOG × ELOG etc.

B_STAND = Dummy variable: Whether respondents felt bringing water quality up to bathing standards was worthwhile, irrespective of cost; YES = 1; NO = 0;

D_STAND = Dummy variable: Whether respondents felt bringing water quality up to drinking standards was worthwhile, irrespective of cost; YES = 1; NO = 0;

CHENNAI, DELHI etc. = Dummy variables: Whether the respondent resided in that particular city: Yes = 1; No = 0.

where
TVM is Trivandram
BLE is Bangalore
KLA is Calcutta
HYV is Hyderabad

In each case the regressions were analysed using Ordinary Least Squares (OLS) with corrections made for heteroscedasticity when it was detected (using a Breusch–Pagan Chi-Squared Test). The best results from each of the three specifications tried for each equation are presented in Tables 5.2 to 5.4, and Table 5.6 gives the mean WTP estimated from each equation. The best results in each case were from from the Translog specifications of the WTP functions.

There are several points of interest in the estimates of the three WTP functions. First, the regression equations have low Adjusted R^2 values. This usually indicates that the estimated equation is not a good fit to the data.[9] However, it must be borne in mind that the reported estimation results are for cross-section data, which do not usually give as high Adjusted R^2 values as time-series data. Still, the generally acceptable range being 20 per cent and above, these regressions do not represent a particularly good fit to the data.

Second, very few of the variables estimated are statistically significant (indicated by the low values of the relevant t-statistic, and by the absence of the asterisks that denote significance at either the 1 per cent, the 5 per cent, or the 10 per cent error levels). In particular, the income variable is not statistically significant, and does not have the correct

[9]Note however, that the F-statistics are strongly significant in all cases. The F-statistic reports on the overall significance of the regression equation. It tests the (null) hypothesis that the estimated coefficients are equal to zero. Any value of the F-statistic greater than 3 indicates that this hypothesis is strongly rejected and that the estimated regression is statistically significant.

TABLE 5.2

Estimation Results for Willingness-to-pay for Best or Bathing Quality

Dependent variable:	BLOG	No. of observations:	817
R²:	20.9934	Adjusted R²:	18.4964
F-statistic (zero slopes):	8.76810	Econometric Package:	TSP 4.3

Variable	Coefficient	t-statistic
CONSTANT	22.187500	1.22247
ILOG	−0.400438	−0.28322
ALOG	−14.502600*	−3.27280
SLOG	6.121680†	1.92340
ELOG	4.627910	0.60973
ILOGI	0.180824	1.53299
ILOGA	−0.014331	−0.08077
ILOGS	0.201127	1.10851
ILOGE	−0.482672	−1.09420
ALOGA	2.817740*	3.28219
ALOGS	−0.449104	−1.30465
ALOGE	1.596020	1.52355
SLOGS	0.516816	1.14340
SLOGE	−2.488360*	−2.69701
ELOGE	−0.648300	−0.56828
VISIT	0.162346	1.61597
INF_CL	0.327139†	2.67511
B_STAND	0.120155	0.65704
D_STAND	0.124250	1.45915
CHENNAI	0.581799*	3.42109
DELHI	−0.292218†	−2.15646
TVM	−0.103664	−0.52814
BLE	0.098207	0.59785
BARODA	0.216923	1.42000
KLA	−0.593361*	−4.51252
HYV	−0.022829	−0.11628

Notes: Standard Errors are heteroscedastic-consistent (HCTYPE = 2).
*denotes significance at the 1 per cent error level.
†denotes significance at the 5 per cent error level.
‡denotes significance at the 10 per cent error level.

TABLE 5.3

Estimation Results for Willingness-to-pay for Current or 1995 Quality

Dependent variable:	CLOG	No. of observations:	817
R^2:	18.2658	Adjusted R^2:	15.6825
F-statistic (zero slopes):	7.07084	Econometric package:	TSP 4.3

Variable	Coefficient	t-statistic
CONSTANT	11.999700	0.64047
ILOG	−0.460125	−0.30164
ALOG	−9.628480†	−2.31354
SLOG	6.476950‡	1.93079
ELOG	4.832170	0.60945
ILOGI	0.193817	1.57214
ILOGA	−0.085806	−0.47969
ILOGS	0.199986	1.07353
ILOGE	−0.432471	−0.93856
ALOGA	1.987450†	2.34318
ALOGS	0.518885	1.53184
ALOGE	−1.274220	−1.24617
SLOGS	0.424339	0.92987
SLOGE	−2.495280†	−2.64351
ELOGE	−0.408042	−0.34718
VISIT	0.154259	1.43517
INF_CL	0.271453†	2.15981
B_STAND	0.137684	0.75519
D_STAND	0.069910	0.79524
CHENNAI	0.755448*	4.20909
DELHI	−0.284111‡	−1.85809
TVM	0.056284	0.27586
BLE	0.243771	1.45623
BARODA	0.097441	0.59576
KLA	−0.558114*	−4.22443
HYV	0.121305	0.61557

Notes: Standard Errors are heteroscedastic-consistent (HCTYPE = 2).
    *denotes significance at the 1 per cent error level.
    †denotes significance at the 5 per cent error level.
    ‡denotes significance at the 10 per cent error level.

TABLE 5.4
Estimation Results for Willingness-to-pay for 1995 Quality

Dependent variable: PLOG   No. of observations: 603
R^2:   15.5058   Adjusted R^2:   11.8449
F-statistic (zero slopes): 4.23548 Econometric package: TSP 4.3

Variable	Coefficient	t-statistic
CONSTANT	26.043700	1.18962
ILOG	1.526550	0.78344
ALOG	−16.724600*	−3.46321
SLOG	9.843610†	2.64280
ELOG	−4.794390	−0.54159
ILOGI	0.000870	0.00623
ILOGA	−0.090524	−0.38705
ILOGS	−0.067364	−0.31384
ILOGE	−0.303969	−0.59221
ALOGA	2.624700†	2.61340
ALOGS	−0.597472	−1.40539
ALOGE	3.040770†	2.41785
SLOGS	−0.120832	−0.22846
SLOGE	−2.413130†	−2.23986
ELOGE	−0.087979	−0.06357
VISIT	0.346452†	0.28377
INF_CL	0.344856†	2.32958
B_STAND	0.156320	0.72446
D_STAND	0.177749‡	1.69830
CHENNAI	0.748247*	3.02631
DELHI	0.050339	0.26795
TVM	0.188921	0.89175
BLE	0.389811‡	1.89416
BARODA	0.370622	1.37254
KLA	−0.355886†	−2.37680
HYV	0.428433	1.37432

Notes: Standard Errors are heteroscedastic-consistent (HCTYPE = 2).
  *denotes significance at the 1 per cent error level.
  †denotes significance at the 5 per cent error level.
  ‡denotes significance at the 10 per cent error level.

(positive) sign in all cases. Since income is expected to be positively related to the willingness-to-pay, and significantly so, this indicates further that the estimation does not represent a good fit to the data.

Third, the variable representing the age of the respondents is strongly negatively related to the willingness-to-pay, albeit in a regression which may not have a very good overall fit with the data. This would indicate that younger respondents tend to give higher values for non-user benefits. However, since the cross-term (ALOGA) in the Translog specification is strongly positive in all cases, there could be a non-linear relationship between age and willingness-to-pay. The best measure of the responsiveness of WTP to the variable AGE, nevertheless, is the elasticity of WTP to AGE. The elasticity in the case of the first regression estimation (for Best or Bathing Quality in Table 5.2) is –10.5523, indicating that if age goes up by 1 per cent, willingness-to-pay falls by 10.55 per cent.

The dummy variables show that prior information about the Ganga Action Plan increase the WTP of the respondents, but the view that either bathing or drinking quality water is a worthwhile objective irrespective of cost does not have a statistically significant impact on subsequent WTP offers. In terms of the city dummy variables, the results show that respondents from Chennai tend to give higher estimates of WTP than those from other cities, while the Delhi and Kanpur–Lucknow–Allahabad respondents tend to give lower estimates.

Replacing the regressors (right-hand side variables) with their mean values and exponentiating the right-hand side of the regression equations will yield the mean values of the left-hand side dependent variables that we are interested in, namely, BESTMAX, CURRMAX, and PASTMAX. The results are given in Table 5.6.

There is, however, another way of deriving the mean willingness-to-pay, using water quality as an explicit variable in the right-hand side of the regression equation. This is described in the following model.

ECONOMETRIC MODEL WITH WATER QUALITY

Two new variables are created in this model. First, a variable called QUALITY is created as an index of river water quality. This index is calculated using the statistics on BOD (Biological Oxygen Demand) measured at different points along the Ganga in the years 1987 and 1996 (the two years closest to the years chosen, for which complete data was available). This weighted average was suitably calibrated to increase with a fall in BOD, or an increase in river water quality (see Appendix 5.2 for details of the calculation of this index). The values taken by the index in the three scenarios are:

Best Quality = 100.00;
1995 Quality = 48.63;
1985 Quality = 31.46.

The second new variable is the dependent variable in the regression equation, Log(BID) (or LBID), where

BID is BESTMAX   when QUALITY is 100.00;
BID is CURRMAX   when QUALITY is 48.63;
BID is PASTMAX   when QUALITY is 31.46.

The sample used in the regression, thus, is three-times the All Non-users sample used in the previous regressions. The actual specification of the regression equation (in Cobb–Douglas logarithmic linear form) is:

$$\text{LBID} = \alpha + \alpha_Q \text{ QLOG} + \alpha_I \text{ ILOG} + \alpha_A \text{ ALOG} + \alpha_S \text{ SLOG}$$
$$+ \alpha_E \text{ ELOG} + \beta_1 \text{ VISIT} + \beta_2 \text{ INF\_CL} + \beta_3 \text{ B\_STAND}$$
$$+ \beta_4 \text{ D\_STAND} + \delta_1 \text{ CHENNAI} + \delta_2 \text{ DELHI} + \delta_3 \text{ TVM}$$
$$+ \delta_4 \text{ BLE} + \delta_5 \text{ BARODA} + \delta_6 \text{ KLA} + \delta_7 \text{ HYV} + \varepsilon; \quad (5.3)$$

where

QLOG	= Log of QUALITY as defined above
ILOG (or Log INCOME)	= Log (per capita annual household income from all sources)
ALOG (or Log AGE)	= Log (Age of the respondent)
SLOG (or Log SIZE)	= Log (Size of the household: with members below the age of 18 converted to adult equivalent units)
ELOG (or Log EDU)	= Log (Education level of the respondent, in years)
VISIT	= Dummy variable: Whether respondents had visited the Ganga in the past 10 years: YES = 1; NO = 0
INF_CL	= Dummy variable: Whether or not respondents had heard about the Ganga Action Plan; YES = 1; NO = 0
B_STAND	= Dummy variable: Whether respondents felt bringing water quality up to bathing standards was worthwhile, irrespective of cost; YES = 1; NO = 0

D_STAND = Dummy variable: Whether respondents felt bringing water quality up to drinking standards was worthwhile, irrespective of cost; YES = 1; NO = 0

CHENNAI, DELHI etc. = Dummy variables: Whether the respondent resided in that particular city; YES = 1; NO = 0.

The results from this regression are detailed in Table 5.5.

TABLE 5.5

Estimation Results for Model with Quality as a Regressor

Dependent variable:	LBID	No. of observations:	2237
R^2:	33.4308	Adjusted R^2:	32.95110
F-statistic (zero slopes):	69.6797	Econometric package:	TSP 4.3

Variable	Coefficient	t-statistic
CONSTANT	−3.931600*	−5.08746
QLOG	1.473850*	26.77850
ILOG	0.285526*	6.17231
ALOG	−0.373782*	−4.24718
SLOG	0.236715*	2.99498
ELOG	0.352119‡	1.61129
VISIT	0.226198*	3.35536
INF_CL	0.292159*	3.83297
B_STAND	0.176796	1.61140
D_STAND	0.125844†	2.35420
CHENNAI	0.784868*	6.78864
DELHI	−0.173690†	−1.91892
TVM	0.098726	0.84003
BLE	0.207553†	2.01479
BARODA	0.218628†	2.03979
KLA	−0.540680*	−6.77557
HYV	0.253662†	2.02352

Notes: Standard Errors are heteroscedastic-consistent (HCTYPE = 2).
  *denotes significance at the 1% error level.
  †denotes significance at the 5% error level.
  ‡denotes significance at the 10% error level.

These results have several noteworthy features. First, the value of the Adjusted R^2 is significantly higher than those obtained for the earlier regressions (from the model which did not take river water quality as an independent variable or regressor). The value for the current regression is well within the acceptable range for cross-sectional data.

Second, almost all the variables are highly statistically significant (indicated by the high t-ratios). Along with the fact that the Adjusted R^2 is high (as well as the F-ratio, which represents the overall significance of the regression), these high t-ratios make the interpretation of the results more meaningful.

Third, all the variables have the expected sign; income is expected to be positively related to WTP, as are the size of the household and the educational level. The educational level may also be interpreted as a proxy for the environmental awareness of the respondent, although this relationship need not obtain in general.

The elasticity of WTP with respect to income is the value of the estimated coefficient of the variable ILOG (the natural logarithm of annual gross household income from all sources), which in this case is 0.2855. This means that if income rises by one per cent, WTP for better water quality in the Ganga may be expected to rise by 0.2855 per cent. Although this is perhaps low in comparison to the comparable elasticities (in the context of improvements in freshwater quality) in developed countries, this is not unusual, especially for non-user benefits valued in a developing country.

The variable AGE is negatively correlated with income, and significantly so, confirming the result from the earlier regression that younger respondents tended to give higher values of WTP for water quality improvements in the Ganga.

Replacing all the right-hand side variables by their (sample) mean values and setting the QUALITY index equal to 100 gives the estimated household WTP for best or bathing quality, while setting the QUALITY index equal to 48.63 gives the estimated household WTP for current quality (1995), and setting it at 31.64 gives the estimated household WTP for past quality (1985).

The estimated household WTP for each of these water quality levels, from using three different methods, is presented in the following section.

ESTIMATES OF MEAN WILLINGNESS-TO-PAY AND AVERAGE
VALUATION OF INCREASES IN RIVER QUALITY

Table 5.6 details the three household willingness-to-pay (WTP) estimates that result from using the sample median, the estimated mean from the econometric model without water quality, and estimated mean from the econometric model with water quality, for different levels of water quality.

TABLE 5.6

Mean Willingness-to-pay for All Non-users

(Rupees per household per annum, at 1995–6 prices)

Basis for household WTP calculation	Levels of river water quality			
	Best quality with GAP	1995 quality with GAP	1985 quality with GAP	1995 quality without GAP
1. Sample median	500.00	200.00	100.00	
2. Estimated mean (model without quality)	533.02	217.79	91.64	
3. Estimated mean (model with quality)	557.94	192.81	101.48	97.51

The level of non-user benefit of achieving bathing quality (the ultimate GAP objective) to an urban literate household[10] ranges between the sample median figure of Rs 500 per annum and the value of the estimated mean from the econometric model with water quality included as an explanatory variable, Rs 557.94 per annum.

The value of non-user benefits from the current level (1995) of water quality ranges similarly from Rs 192.81 to Rs 217.79 per household per annum, while that of non-user benefits from the past quality (1985) ranges from Rs 91.64 to Rs 101.48 per household per annum.

The econometric model with the quality variable makes possible a useful further calculation: the non-user benefits from the scenario which would have existed currently (in 1995) *without* GAP, with the Index of River Water Quality (see Appendix 5.2) set at 30.62 (the value the Index takes with BOD levels that simulations on the Water Quality Model indicate would obtain in 1995 if the Ganga Action Plan had not started). Setting the QUALITY variable equal to 30.62 in the estimated regression equation yields the level of annual non-user benefits each household could expect to receive in the absence of the Ganga Action Plan, which is the number in the bottom right-hand corner of Table 5.6, i.e. Rs 97.51.

---

[10]When reference is made to a household being 'educated' or 'literate', it is simply to indicate that the head of the household or the respondent from the household (above the age of 21 years and responsible enough to make financial decisions for the household) is either educated or literate.

In addition to these household values, incremental values of WTP can also be generated. For instance, the difference between the WTP for 1995 Quality and 1985 Quality gives the value of benefits (per household per annum) from the improvements in water quality carried out between 1985 and 1995. Similarly, the difference between the WTP for Best or Bathing Quality and 1995 Quality, gives the value of potential benefits from cleaning the river up to uniform bathing quality throughout the river. These incremental value calculations presented in Table 5.7 are based on for the estimates of household willingness-to-pay given in Table 5.6.

TABLE 5.7

Willingness-to-pay for Improvements in River Water
Quality by All Non-users

(Rupees per household at 1995–6 prices per annum)

Changes in water quality levels	Basis for household WTP calculation		
	Sample median	Estimated mean	
		Model without quality	Model with quality
1. 1985 to bathing quality	400.00	441.38	455.42
2. 1995 to bathing quality	300.00	315.22	365.13
3. 1985 to 1995 quality	100.00	126.15	91.34
4. Simulated to actual (1995) quality	–	–	95.30
5. Simulated (1995) to bathing quality	–	–	460.43

Table 5.7 shows that the value of the benefits to an average urban Indian literate and employed household of improving river water quality from the 1985 level to bathing quality (the ultimate objective of the GAP) ranges between Rs 400 and Rs 455.42 per annum.

The estimated benefit of raising water quality from the current (1995) level to bathing quality ranges from Rs 300 to Rs 365.13 per household per annum, and that of increasing water quality in the Ganga from 1985 to the present (1995) ranges between Rs 91.34 and Rs 126.15. This last value represents one interpretation of the benefit to the average literate and employed urban Indian household of having the Ganga Action Plan.

Another measure of the estimated benefit of the GAP is the difference between the benefits that would accrue to households in 1995 had the GAP *not* been implemented (the 'without GAP scenario') and those that have actually accrued in 1995 as a result of the GAP. This calculation is only possible when using the model with quality included as an explanatory variable, and the figure is Rs 95.30 per household per annum.

Finally, this estimation of benefits gives another measure of the total benefit the GAP can be expected to have, once the objective of uniform bathing quality is achieved. This is the difference between the benefit that would have accrued in 1995 in the absence of the GAP and the total benefit of achieving bathing quality, which is the number in the bottom-right hand corner of Table 5.7, that is, Rs 460.43 per annum per household.

From each of these household WTPs, a population estimate has to be computed in order to generate total WTP for each level of water quality and for the incremental benefits calculated. This is done in the following section.

AGGREGATE WILLINGNESS-TO-PAY FOR WATER QUALITY LEVELS AND IMPROVEMENTS

There are two stages in the extrapolation of household willingness-to-pay estimates to the population: first, calculating the size of the target population sampled; and second, multiplying the mean willingness-to-pay by this population size. Keeping in mind that the population sampled is the urban literate population in cities with a population of over 1 million households, the steps detailed in Table 5.8 yield the desired total population size.

TABLE 5.8

The Total Population Sampled: Urban Literate Households in India

Number of urban households in India:	41.418 million
Average household size (all India urban population):	5.34
Urban literacy rate (all-India national average):	66.00 %
Number of Indian cities with populations above 1 million:	23
Total population in these cities:	70.661 million
Total number of households in these 23 cities: (70,661,000/5.34)	13.232 million
Total number of literate households in these 23 cities: (13,232,397 × 0.66)	8.733 million

Using this total population figure of 8.733 million households, the household willingness-to-pay (in terms of rupees per household per annum) can be aggregated into a population or total willingness-to-pay. These figures, for each set of household willingness-to-pay estimates, generated above in Tables 5.6 and 5.7 are presented in Tables 5.9 and 5.10.

Table 5.9 presents the aggregated value of benefits to urban literate and employed households for different levels of water quality, including the simulated 1995 quality in the absence of the Ganga Action Plan. This table shows that the value of aggregate benefits to the urban employed and literate population in India (for 23 cities with a population greater than 1 million) from a Ganga that is cleaned up to bathing quality ranges between Rs 4366.69 million and Rs 4872.694 million per annum.

TABLE 5.9

Aggregate Willingness-to-pay for Ganges Water Quality Levels

(Rs million per annum)

Level of water quality	Basis for household WTP calculation		
	Sample median	*Estimated mean*	
		Model without quality	*Model with quality*
Bathing quality	4366.7	4655.0	4872.7
Current quality (1995)	1746.7	1902.0	1683.9
Past quality (1985)	873.3	800.3	851.6
Simulated 1995 quality without GAP	–	–	886.2

The value of aggregate benefits to this population from the current quality (1995) of Ganga ranges from Rs 1683.9 million to Rs 1902 million per annum, while that from the 1985 quality of Ganga (that is, before the GAP) is between Rs 800.3 million and Rs 873.3 million per annum. An additional figure is available from the model with quality included as an explanatory variable, the aggregate value of benefits from the level of water quality that would have existed in the Ganga *in the absence* of the GAP. This value is Rs 886.2 million per annum.

Table 5.10 shows the aggregated value of the benefits of changing the level of water quality in the Ganga. If the quality were improved to bathing quality (the ultimate GAP objective), the incremental benefit from the period of the inception of GAP would be between Rs 3493.4 million and Rs 4021.1 million per annum. This is one interpretation of the overall benefit from the GAP (Phases I and II) to the urban employed and literate population in the 23 major cities (with a population of over 1 million) in the country.

TABLE 5.10

Aggregate Willingness-to-pay for Changes in
Ganges Water Quality Levels

(Rupees million per annum)

Changes in water quality levels	Basis for household WTP calculation		
	Sample median		Estimated mean
	Model without quality	Model with quality	
1985 to bathing quality	3493.4	3854.7	4021.1
1995 to bathing quality	2620.0	27523.0	3188.8
1985 to 1995 quality	873.3	1101.7	797.7
Simulated to actual (1995) quality	–	–	832.3
Simulated (1995) to bathing quality	–	–	3986.5

If river water quality improved from the current (1995) quality to bathing quality, the value of additional non-user benefits generated ranges from Rs 2620 million to Rs 3188.8 million per annum.

The value of additional non-user benefits to urban literate and employed households of river water quality improvements from 1985 to 1995, the period of interest in the present evaluation study, ranges from Rs 797.7 million to Rs 1101.7 million per annum.

Further, the benefits of having the GAP today could be interpreted as the value of the aggregate incremental benefits, between the value given for actual current (1995) quality and that of simulated current quality (1995) (that is the value had the GAP not taken place), which is Rs 832.3 million per annum.

Note that there are two interpretations of the benefit of having GAP (Phase I) today: (1) the incremental benefits from 1985 to 1995; and (2) the difference between the benefits from the current (1995) quality (with the GAP) and those from the simulated current (1995) quality (without the GAP). Using the results from the econometric model with quality included as an explanatory variable, the value for the first is Rs 797.7 million and that of the second is Rs 832.3 million. The latter is thus 4.32 per cent higher than the value based on the difference between actual 1985 and 1995 water qualities, reflecting the fact that 1995 quality without the GAP is worse than actual 1985 quality.

Lastly, the benefits of having the GAP (Phases I and II) may be interpreted as the value of aggregate incremental benefits from the water quality simulated for 1995 and that from uniform bathing quality in the river, the figure in the bottom right-hand corner of the Table 5.10, Rs 3986.5 million per annum.

Note that again there are two ways of interpreting the benefit of having the GAP (Phases I and II): (1) as the difference between the past (1985) value of benefits and those from achieving uniform bathing quality, that is Rs 4021.10 million per annum and (2) as the difference between the simulated (1995) value of benefits and the benefits from having uniform bathing quality water in the river, which is the number in the bottom right-hand corner of the Table 5.10, that is Rs 3986.5 million per annum. The former is less than 1 per cent higher than the latter.

CONCLUSION

This study provides further support to the view that the Contingent Valuation Method can be used successfully to measure non-user benefits from environmental resources in developing countries. The implementation of carefully designed questionnaires to survey sample households in eleven major cities in India evoked unexpectedly good responses about the households' valuation of non-user benefits for an important water resource in the country. A part of this response may, no doubt, be because the sample consisted of households from the literate urban population in the country, with a capacity to appreciate environmental concerns in the country and, in particular, the concern about the control of pollution in a river of major economic and religious significance like the Ganga. Apart from providing data for estimating the household valuation of non-user benefits, the survey can also be said to have performed the important social function of increasing environmental awareness of the urban population in India.

The surveyed urban households revealed consistent preferences for the quality of Ganga in the three scenarios presented—(1) 1985 quality (that is, before the GAP), (2) current quality, (1995), and (3) bathing quality (the ultimate objective of the GAP)—with the household valuations rising with the increased quality of the river. Data also showed that their preferences were also consistent in that valuations increased with income, size, and environmental awareness of the household.

The estimation results from the econometric model with water quality included as an explanatory variable had a good measure of fit to the data and all the variables of the correct sign (denoting consistent preferences) and were highly statistically significant.

The main results from this model show that for the 8.7 million households from the urban literate population in 23 major cities (with a population of more than 1 million) in India, the average annual willingness-to-pay (WTP) for bathing quality water in the Ganga is Rs 557.94 per household per annum, which yields an aggregate figure of Rs 4872.7 million per annum.

The average annual WTP for current quality (1995) in the Ganga is Rs 192.81 per household per annum, which aggregates to Rs 1683.9 million per annum.

The value of non-user benefits for past quality, (1985) that is, the quality that existed before the GAP was initiated, is Rs 101.48 per annum per household, and totals Rs 851.6 million per annum.

The econometric model with water quality makes a further calculation possible. Using an Index of River Water Quality developed on the basis of data from the National Rivers Conservation Directorate, Ministry of Environment and Forests, and simulations on the Water Quality Model done by the Industrial Toxicology Research Centre, Lucknow, the non-user benefits of the water quality that would have existed in 1995 *in the absence of* the GAP can be worked out. This figure is Rs 97.51 per household per annum, on average, and Rs 886.2 million per annum in total.

Using these mean WTP values for non-user benefits, the benefit of having the GAP today can be estimated in two ways: (1) the incremental benefits from 1985 to 1995; and (2) the difference between the benefits from the actual current quality (1995) and those from the simulated current quality (1995, without the GAP). The former is Rs 797.7 million per annum and the latter is Rs 832.3 million per annum, being 4.32 per cent higher than the former, reflecting the fact that water quality in 1995 without the GAP would be worse than the 1985 quality with GAP.

Similarly the benefit of having the GAP (Phases I and II) can be interpreted in two ways: (1) as the difference between the past (1985) value of benefits and those from achieving uniform bathing quality; and (2) as the difference between the simulated value (1995) of benefits and the benefits from having uniform bathing quality water in the river. The former of these amounts to Rs 4021.1 million per annum, while the latter is about 1 per cent lower at Rs 3986.5 million per annum.

6

Measuring User Benefits from Cleaning-up the Ganges

INTRODUCTION

This chapter evaluates the benefits of the GAP to those who normally reside near the river, and who use the river directly, and also to those who may visit the river for pilgrimages or tourism. Such benefits are otherwise called 'user' or 'direct-user' benefits.

We begin by discussing briefly the type of user benefits being surveyed before detailing the questionnaire format used for this survey and then presenting the results of the analysis. These results are compared to estimates of user value from river water quality improvements from other studies in developing countries, before the main points of the analysis of user benefits are drawn together.

USER BENEFITS OF GAP AND THEIR MEASUREMENT

Benefits from the Ganga Action Plan (GAP) that accrue to people who stay near the river or visit the river for pilgrimages, or for tourism, are broadly termed user benefits. These can include benefits from having cleaner river water for irrigation, for direct drinking, washing and bathing, and for industrial use, as well as benefits from having a better river environment (with greater biodiversity) for recreation and aesthetic enjoyment. These direct user benefits can be further divided into user benefits accruing to the urban and rural population, since the type of uses the river is put to can vary across rural and urban areas. The major distinction, of course, will be that rural areas use water for irrigation either directly from the river, or through wells which are recharged by the river. Also, tourism and pilgrimages tend to be concentrated in urban areas.

This study, however, concentrated on measuring only direct-user benefits to urban populations from the GAP. It did not cover non-user benefits that may accrue to direct users, for two reasons. First, non-user values were already being picked up by the survey of non-user benefits (reported in the previous chapter). Second, there is a caution in the literature on benefit estimation against trying to pick up user and non-user values *together* (as total economic value) and then trying to separate them into user and non-user components.[1] Finally, for reasons mentioned earlier, the survey concentrated on literate and employed populations in cities along the Ganges, though a part of the sample covered pilgrims and tourists to the major religious sites. Thus, the environmental amenity being valued and the target population are: direct user benefits of the GAP to the urban literate, and the employed population living along the river (as well as tourists and pilgrims at the major religious sites) respectively. The method of measurement is contingent valuation.

CONTINGENT VALUATION OF USER BENEFITS OF THE GAP

SALIENT FEATURES OF THE CONTINGENT VALUATION SURVEY

As in the case of the contingent valuation survey of non-user benefits, the survey of user benefits had to address a *post hoc* evaluation of benefits from the GAP, of water quality improvements from 1985 (the year of inception of the GAP) to 1995. The approach taken was similar to that adopted in the non-user survey (reported in Chapter 5 above), namely, to ask respondents for an explicit valuation of three different levels of water quality: (1) as it was in 1985, before the GAP; (2) as it was in 1995, ten years after the GAP was initiated; and (3) as it was projected to be after the completion of the GAP (Phases I and II). Water quality maps (See Figures 1.2, 1.3, and 1.4) were used once again and respondents were told to assume, in each case, that water quality was as shown in the map and that no further changes were to be expected. This allowed, as in the previous case, for the calculation of incremental differences in benefits received.

The choice of measure of benefit was, once again, consumers' com-

---

[1] For instance, Mitchell and Carson (1989) write: 'While we believe CV surveys are capable of measuring benefits that include a non-use dimension, we are less optimistic about their ability to obtain meaningful estimates of separate component values' (p. 67).

pensating surplus, to be captured by using a willingness-to-pay format. The elicitation format used was also an open-ended bidding game, using a variant of the 'Payment Card', where respondents were asked two follow-up questions to their initial (non-zero) bid (for further details see previous chapter).

THE QUESTIONNAIRE

The final questionnaire used (which is similar to the questionnaire used for non-user benefit survey) had the following features:

Information about Gap

Although those living near the river could be assumed to have a greater knowledge and familiarity with the GAP, pre-testing showed that there were significant gaps in the detailed knowledge of the size of the pollution problem and the various steps being taken by the GAP to tackle the problem along the entire stretch of the river.

The first section of the questionnaire briefly summarized the basic causes of pollution in the river, pointing out that domestic sewage from large cities was the largest cause of pollution (and not industrial pollution, as is commonly understood), and then gave a short account of the activities of the GAP, ranging from setting up sewage treatment plants (to combat sewage-based pollution) to constructing electric crematoria (to control the immersion of unburnt or partly-burnt corpses in the river). The three maps of water quality maps were also shown at this stage.

There were short questions about whether the respondent had visited any other sites along the river in the last ten years, about the ways in which the respondent used or expected to use the river, and whether or not the respondent felt it was worthwhile to improve the water quality for bathing (and then for drinking), irrespective of the cost.

Preference Elicitation

This section carried questions aimed at eliciting consumer preferences for action to clean the river. They were designed to make the respondent think about the responsibility of action to clean the river. The concluding question asked respondents what role they perceived for themselves in the fight against pollution, irrespective of the nature of government action to that end. This led on to the value elicitation questions in the next section.

Value Elicitation

The methodology was similar to that followed in the non-user survey, and so, only the outline of each component of this section are given below.

DETAILING THE HYPOTHETICAL SCENARIO: Respondents were initially given examples of private goods (willingness to pay the cost of a pen in return for the benefits of its use) to illustrate the link between willingness-to-pay and benefits received. Respondents were then asked to consider river water quality as a sort of public good, and given a rough idea of the benefits accruing from various other government-provided public goods, like educational facilities, electricity and power, health facilities and transport, as the amount the government spent per year per household to provide these various public goods, as shown on a card.

SPECIFYING THE BUDGET CONSTRAINT: Apart from a reminder to keep their income and expenditure constraints in mind, respondents were also told to think of user benefits from a clean-up of the river in the same manner as they normally 'valued' benefits from health; while one could consider the benefits of good health as 'priceless' or 'infinite', in practice, one could not spend more than one's income on health facilities. And even then, one spent only a fraction of one's income on them. The intention, as in the case of the survey of non-user benefits, was to try and avoid impossibly high estimates that may be expected in the context of a river with considerable religious and symbolic importance.

In addition, all respondents who answered 'No' to the Respondent Evaluation Question: 'Do you feel you may actually be required to pay for water quality improvements in the Ganges?', were dropped from the sample, the reason being that those who did not feel that their answers had to be constrained by considerations of actual payment may have deliberately overstated their willingness-to-pay.

THE CHOICE OF ELICITATION FORMAT: As in the case of the survey of non-user benefits, a variant of the open-ended bidding game format backed by a version of a 'Payment Card' in a multiple bid format was used, with an initial bid, a second bid, and a final bid. The first bid was open-ended, but based on the Payment Card, followed by a second, take-it-or-leave-it question based on the initial bid, and a final question asked for the maximum they would pay. The response to this last question was taken to be the WTP value. This procedure was followed for the first scenario, that is, best or bathing quality, but only one question, asking for the maximum WTP, was used for the second and third scenarios (1995 and 1985 levels of water quality).

POSING THE VALUATION QUESTION: Value elicitation questions were posed, once again, as explicit valuation problems. Respondents were shown each map of water quality and told explicitly to assume that the water quality shown actually obtained at the present, *and that they should expect no further changes in water quality*. They were then asked to evaluate the benefits they would receive as direct users if water in the river *stayed* at that particular quality level.

Thus, respondents were paying to enjoy the benefits they would receive in the hypothetical situation that the river water quality was at that particular level. This also helped avoid any tendency on the part of the respondents to value worse water quality higher, since they imagined they would have to pay more to clean up a dirtier river.

THE CHOICE OF THE PAYMENT VEHICLE: After considering several alternatives at the pre-testing stage, payment to a reputable charitable organization was chosen as the payment vehicle.

Socio-economic details

This section of the questionnaire collected information on socioeconomic variables to be used in the regression estimation of the valuation function. Apart from the name of the respondent and the address, information was collected on age, educational level (Table 6.1), occupation, size of the household (both above and below 18 years), and gross annual household income from all sources. The last was posed alternatively in the form of bands of possible incomes, which the respondent could choose. Given the proclivity of underestimating income in the country, the subsequent analysis took the upper limit of the income class chosen, as representative of the income of the respondent.

Data on the gender of the head of the household showed that around 12 per cent of those surveyed were female-headed households. No attempt was made to analyse data for the sub-sample of female-headed households.

TABLE 6.1
Frequency Distribution of User Samples by Education Levels

Education Level	User Sample (%)
Up to class XII	21.6
Graduate (B.A.)	33.7
Postgraduate (M.A.)	28.2
Professional qualification	13.1
Other qualification	3.3

Respondent's Evaluation

Several questions were asked in this section as a follow-up to the questions posed in the previous sections. In particular, they asked for the respondent's evaluation of the clarity of the enumerator's presentation of the questionnaire, the extent to which the respondent believed the answers provided would influence policy and affect clean-up operations, and whether or not the respondent believed actual payments would be asked for. These helped to screen out questionnaires whose responses might be biased for a lack of clarity of exposition by the enumerator.

Enumerator's Evaluation and Declaration

Two questions were put to the enumerators at the end of the interview, to ascertain the attitude and capability of the respondent to answer the elicitation questions. They asked the enumerators to describe the level of perceived effort by the respondents to focus on the hypothetical scenario and to understand the context and purpose of the elicitation questions. Questionnaires where enumerators felt respondents either did not pay sufficient attention to the valuation questions or did not understand the questions were subsequently eliminated from the sample.

Further, enumerators were asked to sign a declaration stating that the interview had been conducted honestly, and in accordance with the instructions they had received during their training. The basic purpose of these questions was to instil a sense of responsibility in the enumerators in exerting their utmost to elicit accurate responses.

THE SAMPLING

The sample was limited to residents, tourists, and pilgrims (at the bathing ghats) in the major cities along the river. Three cities were sampled; Allahabad, Varanasi, and Calcutta, with a total target sample size of 300 per city, giving a total target sample of 900. The final sample size was 845, the rest being protest bids, incompletely filled-in questionnaires, or questionnaires where respondents had either not understood the hypothetical scenario or did not expect to actually pay for improvements in water quality according to their stated willingness-to-pay.

Within each city, the sampling scheme was designed to cover residential areas within half a kilometre on either side of the river and all major pilgrimage centres (bathing ghats), with control for geographical spread and income categories. Again, at least as far as the residential sample was concerned, only the literate and the employed were chosen from among the urban population living within a range of half a kilometre from the river. Reviewers of the draft report have questioned these limits, and have argued that user benefits could be spread wider than 0.5 kilometers on each side of the river. Some of these benefits are picked up by the agricultural and health benefits but it excludes visitors to the sites from further afield. We acknowledge that this assumption has probably resulted in some underestimation of the benefits, which, unfortunately cannot be corrected at this stage.

The choice of enumerators was again restricted to college and University students, who were each paid Rs 25 per fully filled-in questionnaire, plus Rs 10 for local travel costs. The following section presents the analysis of the data and the results.

ESTIMATION RESULTS

There are three sets of results. First, the calculation of the mean willingness-to-pay (WTP). Second, the extrapolation of these average WTP estimates to the population. Third, the calculation of incremental WTP for different sets of changes in water quality.

MEAN WILLINGNESS-TO-PAY

Data from the city samples (Allahabad, Varanasi, and Calcutta) were pooled to get an 'All users sample'. This data was screened to eliminate incomplete questionnaires [including protest (zero) bids] and those which did not reflect the budget constraint. It was also screened to eliminate questionnaires where respondents were assessed by the enumerators as either not understanding the valuation question or not giving it sufficient consideration. This reduced the final sample size by about 30 per cent, from 845 to 576 usable observations. As explained in the case of the contingent valuation survey of non-user benefits, this reduction is not unusually large and includes data screening to ensure that only those households which fully understood the hypothetical scenario and expected to pay for water quality improvements were included in the sample.

Average values for the benefits from three different levels of water quality (Best or Bathing Quality, 1995 Quality and 1985 Quality) are calculated in four different ways: first by taking the sample mean; second, by using the sample median; third, by estimating a regression equation with WTP for each level of water quality regressed on a vector of socio-economic variables and dummy variable, and using the average values of the regressors to generate estimates of the dependent variables; and fourth, by estimating a regression equation with WTP for all levels of water quality regressed on an index of water quality, a vector of socio-economic variables and dummy variables, and using the average values of the regressors and the actual value of the water quality index to generate estimates of the dependent variable. There are thus four separate estimates of mean WTP for each of the three water quality levels, namely

> BESTMAX for Best or Bathing Quality,
> CURRMAX for Current or 1995 Water Quality, and
> PASTMAX for Past or 1985 Water Quality.

In addition to the sample mean, the sample median is used to provide a measure of central tendency in the sample, when the sample is believed to be highly skewed since the median is less influenced by outliers than the arithmetic mean.

ECONOMETRIC MODEL WITHOUT WATER QUALITY

In the regression equation estimated for each level of water quality, the dependent variable is the WTP elicited from respondents for that particular level of water quality. The vector of socio-economic variables on which these dependent variables were then regressed are income, age, education level, and household size, while dummies were added for specific responses given in the questionnaire and also for cities where the respondents stayed (described in detail below). Three functional forms were tried for each equation: the Cobb–Douglas form (in logarithmic-linear form), a modified Cobb–Douglas form (where income and age were entered in squared form), and the Translog functional form. The final equation used in the calculation of the mean value was chosen on the basis of the highest Adjusted R^2 and F-statistics. The full specification of each equation in general form and in Cobb–Douglas form is given below.

General Form

$$\text{WTP} = f(\text{CONSTANT, INCOME, AGE, SIZE, EDU, VISIT,}$$
$$\text{B\_STAND, D\_STAND, CITY1, CITY2, CITY3),} \quad (6.1)$$

where WTP is BESTMAX for best or bathing water quality;
CURRMAX for current or 1995 quality (with GAP); and
PASTMAX for past or 1985 quality.

Cobb-Douglas Logarithmic Linear Form

$$\text{BLOG (or CLOG or PLOG)} = \alpha + \alpha_I \text{ ILOG} + \alpha_A \text{ ALOG} + \alpha_S \text{ SLOG}$$
$$+ \alpha_E \text{ ELOG} + \beta_1 \text{VISIT} + \beta_2 \text{ B\_STAND}$$
$$+ \beta_3 \text{ D\_STAND} + \delta_1 \text{ CIT1Y}$$
$$+ \delta_2 \text{ CITY2} + \delta_3 \text{ CITY3} + \varepsilon, \quad (6.2)$$

where

BLOG (or Log BESTMAX)	= Log (WTP for best quality);
CLOG (or Log CURRMAX)	= Log (WTP for 1995 quality);
PLOG (or Log PASTMAX)	= Log (WTP for 1985 quality);
ILOG (or Log INCOME)	= Log (per capita annual household income from all sources);
ALOG (or Log AGE)	= Log (age of the respondent);
SLOG (or Log SIZE)	= Log (size of the household: with members below the age of 18 years; converted to adult equivalent units);

ELOG (or Log EDU)	= Log (education level of the respondent, in years);[2]
VISIT	= Dummy variable: Whether respondents had visited the Ganga in the past 10 years: YES = 1; NO = 0;
B_STAND	= Dummy variable: Whether respondents felt bringing water quality up to bathing standards was worthwhile, irrespective of cost; YES = 1; NO = 0;
D_STAND	= Dummy variable: Whether respondents felt bringing water quality up to drinking standards was worthwhile, irrespective of cost; YES = 1; NO = 0;
CITY1	= Dummy variables: Whether the respondent resided in the city of Allahabad: Yes = 1; No = 0;
CITY2	= Dummy variables: Whether the respondent resided in the city of Varanasi: Yes = 1; No = 0;
CITY3	= Dummy variables: Whether the respondent resided in the city of Calcutta: Yes = 1; No = 0.

In each case the regressions were analysed using Ordinary Least Squares (OLS) with corrections made for heteroscedasticity when it was detected (using a Breusch–Pagan Chi-squared Test).

The results of each equation (in Translog form) are presented in Tables 6.2 to 6.4, while Table 6.6 gives the mean WTP estimated from each equation.

The best results in each of these cases were from the Translog specification of the WTP functions.[3]

---

[2]For the Translog functional form, the following additional variables were used:

ILOGI = (ILOG × ILOG)/2
ILOGA = ILOG × ALOG
ILOGE = ILOG × ELOG etc.

[3]See footnote 9 in Chapter 5 for a discussion of the use of the Translog specification in the WTP function and of the significance of the adjusted R-squared and the F-ratio in interpreting the 'goodness of fit' of a regression equation.

There are several points of interest in the estimates of each of the three WTP functions. First, the regression equations have low Adjusted R-square (or R^2) values. This usually indicates that the estimated equation is not a good fit to the data. However, it must be borne in mind that the reported estimation results are for cross-section data, which do not usually give as high R^2 values as time-series data. Still, the generally acceptable range being 20 per cent and above, these regressions do not represent a particularly good fit to the data.

Second, very few of the variables estimated are statistically significant (indicated by the low values of the relevant t-statistic, and by the absence of the asterisks that denote significance at either the 1 per cent, the 5 per cent, or the 10 per cent error levels), although the income variable is statistically significant in two regressions, and has the correct (positive) sign in all cases.

Third, the dummy variables show that the view that drinking quality water is a worthwhile objective irrespective of cost had a statistically significant impact on subsequent WTP offers. In terms of the city dummy variables, the results show that respondents from City 2, Varanasi, tend to give lower estimates of WTP than from the other two cities.

Replacing the regressors (right-hand side variables) with their mean values and exponentiating the right-hand side of the regression equations will yield the mean values of the left-hand side dependent variables that are of interest, namely, BESTMAX, CURRMAX, and PASTMAX. The results are given in Table 6.6. There is another way of deriving household WTP, using a regression equation with water quality included as an explanatory variable. This is described in the next section.

ECONOMETRIC MODEL WITH WATER QUALITY

Two new variables are created in this model. First, a variable called QUALITY is created as an index of river water quality. This index is calculated using the statistics on BOD (Biological Oxygen Demand) measured at different points along the Ganga in the years 1987 and 1996 (the two years closest to the years chosen, for which complete data was available). This weighted average was suitably calibrated to increase with a decrease in BOD, or an increase in river water quality (these calculations are reported in detail in Appendix 5.2).

TABLE 6.2

Estimation Results for Willingness-to-pay for Best or Bathing Quality

Dependent variable = BLOG		Observations	= 576	
R^2	= 15.38059	Adjusted R^2	= 12.33125	
F[20, 555]	= 5.043896	Econometric package = LIMDEP 6.0		

Variable	Coefficient	t-ratio	Prob $|t| \geq x$	Mean of X
Constant	−22.31000	−0.896	0.37004	
ILOG	4.70210*	2.575	0.01002	9.9032
ALOG	0.97899	0.169	0.86551	3.7842
SLOG	−0.42064	−0.138	0.89029	1.4626
ELOG	0.38551	0.027	0.97812	2.7293
ILOGI	−0.31633†	−2.459	0.01393	49.5690
ILOGA	0.03340	0.127	0.89893	37.5040
ILOGS	−0.11625	−0.675	0.49957	14.1440
ILOGE	−0.45825	−0.809	0.41867	27.1010
ALOGA	0.48968	0.429	0.66760	7.1956
ALOGS	0.16769	0.423	0.67246	5.5356
ALOGE	−1.23000	−0.879	0.37927	10.3260
SLOGS	0.62075‡	1.785	0.07432	1.2235
SLOGE	0.16299	0.191	0.84870	3.9722
ELOGE	3.49060	0.658	0.51074	3.7360
VISIT	0.02174	0.163	0.87047	0.6684
B_STAND	0.13377	0.836	0.40326	0.93056
D_STAND	0.21604†	2.119	0.03411	0.55729
CITY1	0.07488	0.420	0.67432	−0.23264
CITY2	−0.41026*	−3.444	0.00050	0.29167
CITY3	0.09044	0.621	0.53471	0.18403

Results corrected for heteroscedasticity: Breusch-Pagan chi-squared (DF) = 73.3461 (20).

Note: * denotes significance at the 1 per cent error level.
            † denotes significance at the 5 per cent error level.
            ‡ denotes significance at the 10 per cent error level.

TABLE 6.3

Estimation Results for Willingness-to-pay for Current or 1995 Quality

Dependent variable = CLOG		Observations	= 576
R^2	= 16.50264	Adjusted R^2	= 13.49373
F[20, 555]	= 5.484586	Econometric package = LIMDEP 6.0	

Variable	Coefficient	t-ratio	Prob |t| ≥ x	Mean of X
Constant	−27.51600	−1.019	0.30815	
ILOG	5.62660*	2.739	0.00616	9.9032
ALOG	6.43420	1.034	0.30114	3.7842
SLOG	−0.13860	−0.045	0.96413	1.4626
ELOG	−7.55820	−0.525	0.59946	2.7293
ILOGI	−0.44118*	−3.419	0.00063	49.5690
ILOGA	0.02475	0.094	0.92527	37.5040
ILOGS	−0.27343‡	−1.721	0.08526	14.1440
ILOGE	−0.24523	−0.396	0.69209	27.1010
ALOGA	−0.84481	−0.725	0.46851	7.1956
ALOGS	0.14564	0.372	0.71024	5.5356
ALOGE	−1.32190	−0.870	0.38416	10.3260
SLOGS	0.53761	1.610	0.10737	1.2235
SLOGE	0.68499	0.808	0.41903	3.9722
ELOGE	5.55810	0.987	0.32384	3.7360
VISIT	−0.10349	−0.744	0.45673	0.6684
B_STAND	−0.08214	−0.477	0.63327	0.93056
D_STAND	0.23382†	2.208	0.02726	0.55729
CITY1	−0.00109	−0.006	0.99534	0.23264
CITY2	−0.34891*	−2.886	0.00390	0.29167
CITY3	0.10299	0.640	0.52245	0.18403

Results corrected for heteroscedasticity: Breusch-Pagan chi-squared (DF) = 59.3505 (20).

Note: * denotes significance at the 1 per cent error level.
        † denotes significance at the 5 per cent error level.
        ‡ denotes significance at the 10 per cent error level.

TABLE 6.4

Estimation Results for Willingness-to-pay for 1985 Quality

Dependent variable = PLOG		Observations	= 506
R^2	= 15.85046	Adjusted R^2	= 12.38037
F[20, 485]	= 4.567744	Econometric package = LIMDEP 6.0	

Variable	Coefficient	t-ratio	Prob $|t| \geq x$	Mean of X
Constant	−54.46500‡	−1.861	0.06271	
ILOG	4.33890‡	1.953	0.05087	9.8687
ALOG	9.53710	1.378	0.16811	3.7849
SLOG	4.53370	1.412	0.15783	1.4757
ELOG	9.33070	0.588	0.55660	2.7258
ILOGI	−0.39131*	−2.926	0.00344	49.2210
ILOGA	0.31567	1.129	0.25878	37.3740
ILOGS	−0.28973‡	−1.727	0.08415	14.2280
ILOGE	−0.34576	−0.473	0.63621	26.9750
ALOGA	−0.83047	−0.647	0.51745	7.1983
ALOGS	−0.55527	−1.473	0.14084	5.5896
ALOGE	−3.15650‡	−1.737	0.08238	10.3150
SLOGS	0.11091	−0.320	0.74883	1.2455
SLOGE	0.40150	0.428	0.66867	4.0014
ELOGE	2.50950	0.395	0.69279	3.7269
VISIT	−0.47831*	−2.995	0.00274	0.65415
B_STAND	−0.14992	−0.705	0.48108	0.92885
D_STAND	0.14928	1.223	0.22148	0.57115
CITY1	−0.29921	−1.452	0.14636	0.25296
CITY2	−0.16378	−1.118	0.26337	0.27273
CITY3	−0.18289	−0.908	0.36362	0.18379

Results corrected for heteroscedasticity: Breusch-Pagan chi-squared (DF) = 37.3230 (20).

Note: * denotes significance at the 1 per cent error level.
          † denotes significance at the 5 per cent error level.
          ‡ denotes significance at the 10 per cent error level.

The values taken by the index in the three scenarios are:

> Best Quality = 100.00;
> 1995 Quality = 48.63;
> 1985 Quality = 31.46.

The second new variable is the dependent variable in the regression equation, Log(BID) (or LBID), where

> BID is BESTMAX when QUALITY is 100;
> BID is CURRMAX when QUALITY is 48.63; and
> BID is PASTMAX when QUALITY is 31.46.

The sample used in the regression, thus, is three-times the All Users sample used in the previous regressions. The actual specification of the regression equation (in Cobb-Douglas logarithmic-linear form) is:

General Form

$$BID = f(QUALITY, INCOME, AGE, SIZE, EDU, VISIT, B\_STAND, D\_STAND, CITY1, CITY2, CITY3), \qquad (6.3)$$

Cobb–Douglas (Logarithmic-Linear) Form

$$LBID = \alpha + \alpha_Q\ QLOG + \alpha_I\ ILOG + \alpha_A\ ALOG + \alpha_S\ SLOG$$
$$+ \alpha_E\ ELOG + \beta_1\ VISIT + \beta_2\ B\_STAND + \beta_3\ D\_STAND$$
$$+ \delta_1\ CITY1 + \delta_2\ CITY2 + \delta_3\ CITY3 + \varepsilon \qquad (6.4)$$

where, as before,

LBID	= Log BID
QLOG	= Log (QUALITY);
ILOG (or Log INCOME)	= Log (per capita annual household income from all sources);
ALOG (or Log AGE)	= Log (Age of the respondent);
SLOG (or Log SIZE)	= Log (Size of the household: with members below the age of 18 converted to adult equivalent units);
ELOG (or Log EDU)	= Log (Education level of the respondent, in years);
VISIT	= Dummy variable: Whether respondents had visited the Ganga in the past 10 years: YES = 1; NO = 0;
B_STAND	= Dummy variable: Whether respondents felt raising water quality to bathing standards was worthwhile, irrespective of cost; YES = 1;

D_STAND   = Dummy variable: Whether respondents felt raising water quality to drinking standards was worthwhile, irrespective of cost; YES = 1; NO = 0;

CITY1   = Dummy variable: Whether the respondent resided in the city of Allahabad: Yes = 1; No = 0;

CITY2   = Dummy variable: Whether the respondent resided in the city of Varanasi: Yes = 1; No = 0;

CITY3   = Dummy variable: Whether the respondent resided in the city of Calcutta: Yes = 1; No = 0.

The results from this regression are detailed in Table 6.5.

TABLE 6.5

Estimation Results for Model with Quality as a Regressor

Dependent variable = LBID   Observations   = 1658
R^2 = 33.3227   Adjusted R^2 = 32.8771
F(zero slopes)   = 74.7823   Econometric package = TSP 4.3

Variable	Coefficient	t-statistic
CONSTANT	−8.667210*	−10.17430
QLOG	1.650770*	25.69260
ILOG	0.257506*	5.21845
ALOG	0.220183†	1.84446
SLOG	0.379225*	5.19918
ELOG	1.315010*	5.44689
VISIT	−0.115655	−1.29187
B_STAND	−0.057098	−0.53275
D_STAND	0.207632*	3.19731
CITY1	−0.042507	−0.37118
CITY2	−0.276524*	−3.59519
CITY3	0.043403	0.41013

Standard errors are heteroscedastic-consistent (HCTYPE = 2).

Note: * denotes significance at the 1 per cent error level.

            † denotes significance at the 5 per cent error level.

            ‡ denotes significance at the 10 per cent error level.

These results have several noteworthy features. First, the value of the Adjusted R^2 is significantly higher than those obtained for the earlier regressions from the model which did not take river water quality as an independent variable (regressor). The value for the current regression is well-within the acceptable range for cross-sectional data.

Second, almost all the variables are highly statistically-significant (indicated by the high t-ratios). Along with the fact that the Adjusted R^2 is high (as well as the F-ratio, which represents the overall significance of the regression), these high t-ratios make the interpretation of the results more meaningful.

Third, and quite important, all the variables have the expected signs: income is expected to be positively related to WTP, as are size of the household and the educational level. The educational level may also be interpreted as a proxy for the environmental awareness of the respondent, although this relationship need not obtain in general.

The elasticity of WTP with respect to income is the value of the estimated coefficient of the variable ILOG (the natural logarithm of annual gross household income from all sources), which, in this case, is 0.2575 per cent. This means that if income rises by one per cent, WTP for better water quality in the Ganga may be expected to rise by 0.2575 per cent. Although this is perhaps low in comparison to the comparable elasticities (in the context of improvements in freshwater quality) in developed countries, this is not unusual, especially for non-user benefits.

The variable AGE is positively correlated with income, and significantly so, which is in contrast to the results obtained in a similar regression in the case of the contingent valuation survey for non-user benefits.

Replacing all the right-hand side variables by their mean values and setting the QUALITY index equal to 100 gives the estimated mean WTP for Best or Bathing Quality, while setting the QUALITY index equal to 48.63 gives the estimated mean WTP for Current (1995) Quality, and setting it to 31.64 gives the estimated mean WTP for Past (1985) Quality.

The estimated WTP per household for each of these water quality levels, from using these four different methods is presented in the following section.

ESTIMATES OF PER-HOUSEHOLD WTP FOR LEVELS AND CHANGES IN RIVER WATER QUALITY

Table 6.6 details the three per-household willingness-to-pay estimates that result from using the sample median and the estimated mean from both, the econometric model without water quality and the econometric model with water quality, for each of the three levels of water quality.

TABLE 6.6

Willingness to Pay for All User Households

[Rupees per household per annum (at 1995–6 prices)]

Basis for household WTP calculation	Levels of river water quality			
	Best quality with GAP	1995 quality with GAP	1985 quality with GAP	1995 quality without GAP
Sample median	500.00	200.00	100.00	
Estimated mean (model without quality)	533.31	219.44	75.38	
Estimated mean (model with quality)	581.59	167.23	93.28	71.12

The per-household estimate of direct-user benefits to urban literate households of achieving bathing quality (the ultimate GAP objective) ranges between the sample median figure of Rs 500 per annum, to the estimated mean value from the econometric model with water quality included as an explanatory variable, Rs 581.59 per annum.

The value of user benefits from the current (1995) level of water quality ranges similarly between Rs 167.23 to Rs 219.44 per household per annum, while the user benefits from the past (1985) quality ranges from Rs 75.42 to Rs 100 per household per annum.

The econometric model with the quality variable makes possible a further useful calculation: the user benefits from the scenario which would exist currently (in 1995) without the GAP, with the Index of Water Quality set at 30.62 (the index value for water quality in 1995 if the Ganga Action Plan had not started). Setting the QUALITY variable equal to 30.62 yields the level of annual user benefits each household could expect to receive in the absence of the Ganga Action Plan, which is the number in the bottom right-hand corner of the table: Rs 71.12.

In addition to these values, incremental values of WTP can also be generated. For instance, the difference between the WTP for 1995 Quality and 1985 Quality gives the value of benefits (per household per annum) from the improvements in water quality carried out between 1985 and 1995. Similarly, the difference between the WTP for Best or Bathing Quality and 1995 Quality, gives the value of potential benefits from cleaning the river up to uniform bathing quality throughout the river. These incremental value calculations are presented in Table 6.8 for each of the four sets of estimates of household willingness-to-pay given in Table 6.7.

TABLE 6.7

Willingness-to-Pay for Improvements in
River Water Quality by All Users

[Rupees per household per annum (at 1995–6 prices)]

Change in waters quality level	Basis for household WTP calculation		
	Sample median	Estimated mean	
		Model without quality	Model with quality
1985 to bathing quality	400.00	457.20	488.30
1995 to bathing quality	300.00	313.30	414.35
1985 to 1995 quality	100.00	143.89	73.95
Simulated to actual (1995) quality	–	–	96.11
Simulated (1995) to bathing quality	–	–	510.46

Table 6.8 shows that the value of the benefits to an average urban Indian literate and employed household[4] in cities along the river Ganga, of *improving* river water quality from the 1985 level to bathing quality (the ultimate objective of the GAP) ranges from Rs 400 to Rs 488.30 per annum.

The estimated benefit of raising water quality from the current (1995) level to the bathing quality ranges from Rs 300 to Rs 414.35 per household per annum, and that of increasing water quality in the Ganga from 1985 to the present (1995) ranges from Rs 73.95 to Rs 143.89. This last value is one interpretation of the benefit to the average literate and employed urban Indian household of having the Ganga Action Plan.

[4]When reference it made to a household being 'educated' or 'literate', it is simply to indicate that the head of the household or the respondent from the household (above 21 years and responsible enough to make financial decisions for the household) is either educated or literate.

Another measure of the estimated benefit of having the GAP is the difference between the benefits that would accrue to households in 1995, had the GAP not been implemented (the without GAP scenario) and those that have actually accrued in 1995 as a result of the GAP. This calculation is only possible using the model with quality included as an explanatory variable, and the figure is Rs 96.11 per household per annum.

Finally, this estimation of benefits gives another measure of the total benefit the GAP can be expected to have, once the objective of uniform bathing quality is achieved. This is the difference between the benefit that would have accrued in 1995 in the absence of GAP and the total benefit of achieving bathing quality, which is the number in the bottom-right hand corner of the table: Rs 510.46 per annum per household.

From each of these household WTPs, a population estimate has to be computed in order to generate total WTP for each level of water quality and for the incremental benefits calculated.

AGGREGATE WILLINGNESS-TO-PAY FOR WATER QUALITY LEVELS AND IMPROVEMENTS

There are two stages in the extrapolation of household willingness-to-pay (WTP) estimates to the population: first, calculating the size of the target population sampled; and second, multiplying the household will-ingness-to-pay by the number of households in the target population. Keeping in mind that the population sampled is the urban literate population in cities along the river, the steps detailed in Table 6.8 yield the desired total population size.

TABLE 6.8

The Total Population Sampled—Urban Literate Households in India

Density of Urban Households in Districts (in UP, Bihar, and WB) through which the Ganga flows	22.33
Length of the Ganga River (in km)	2525 km
Area within 0.5 kilometres of the river on either side	2525 km^2
Number of urban households within 0.5 km of the river (household density × area [= $\frac{households}{area}$ × area]= 22.33 × 2525)	56,383.25
Urban literacy rate (average of the rate in the three states)	0.58
Number of literate urban households within 0.5 km of the river (56,383 × 250.58)	37,212.95

Using this total population figure of 37,212,95 households, the household willingness-to-pay figures (in terms of Rupees per household per annum) can be aggregated into a population or total willingness-to-pay (Rupees per annum). These figures, for each set of household willingness-to-pay estimates generated in Tables 6.6 and 6.7 above, are presented in Tables 6.9 and 6.10.

TABLE 6.9

Aggregate Willingness-to-pay for Ganges Water Quality Levels

[Million Rupees per annum (at 1995–6 prices)]

Level of water quality	Basis for household WTP calculation		
	Sample median	Estimated mean	
		Model without quality	Model with quality
Bathing quality	16.4	17.4	19.0
Current (1995) quality	6.5	7.2	5.5
Past (1985) quality	3.3	2.5	3.1
Simulated 1995 quality (without GAP)	–	–	2.3

Table 6.9 presents the aggregated value of benefits to urban literate and employed households for different levels of water quality, including the simulated 1995 quality in the absence of the Ganga Action Plan. This illustrates that the value of aggregate benefits from a Ganga that is cleaned up to bathing quality to the urban employed and literate population in India living in cities along the banks of the river ranges from Rs 16.351 million to Rs 19.019 million per annum.

The value of aggregate benefits to this population from the current (1995) quality of Ganga ranges from Rs 5.468 million to Rs 7.172 million, while that from the 1985 quality of Ganga (that is before the GAP) is between Rs 2.466 million to Rs 3.270 million. An additional figure is available from the model with quality included as an explanatory variable, the aggregate value of benefits from the level of water quality that would have existed in the Ganga in the absence of the GAP. This value is Rs 2.325 million per annum.

TABLE 6. 10

Aggregate Willingness-to-pay for Changes in Ganga Water Quality
Levels by Users

[Rupees million per annum (at 1995–96 prices)]

Change in water quality level	Basis for household WTP calculation		
	Sample median	Estimated mean	
		Model without quality	Model with quality
Users within one km.			
1985 to bathing quality	26	30	32
1995 to bathing quality	20	21	27
1985 to 1995 quality	6	9	5
Simulated to actual (1995) quality	–	–	6
Simulated (1995) to bathing quality	–	–	33
Pilgrims			
1985 to bathing quality	5465	6213	6648
1995 to bathing quality	3800	3956	5377
1985 to 1995 quality	1665	2257	1271
Simulated to actual (1995) quality	–	–	1550
Simulated (1995) to bathing quality	–	–	6927
Total			
1985 to bathing quality	5491	6243	6680
1995 to bathing quality	3820	3977	5404
1985 to 1995 quality	1671	2266	1276
Simulated to actual (1995) quality	–	–	1556
Simulated (1995) to bathing quality	–	–	6960

Table 6.10 shows the aggregated value of the benefits of changing
the level of water quality in the Ganga. If water quality in the river is
improved from the 1985 level to bathing quality (the ultimate GAP objective), the incremental benefit for users within one kilometre from the
period of the inception of GAP would be between Rs 26 million and Rs
32 million. This is one interpretation of the overall benefit from the
GAP (Phases I and II) to the urban employed and literate population
living on the banks of the river Ganga.

If river water quality was improved from the current (1995) quality to bathing quality, the value of additional benefits to users within one kilometre generated ranges between Rs 20 million and Rs 27 million per annum. The corresponding range of values for additional benefits to pilgrims is between Rs 3800 million and Rs 5377 million, giving total values in a range from Rs 3820 million to Rs 5404 million per annum.

The value of additional user benefits to urban literate and employed households living along the river of actual river water quality improvements in the Ganga from 1985 to 1995—the period of interest in the present evaluation study—ranges from Rs 5 million to Rs 9 million per annum. When pilgrims are included, the total range is from Rs 1276 million to Rs 2266 million per annum.

Further, the benefits of having the GAP today could be interpreted as the value of the aggregate incremental benefits between the value given for actual current (1995) quality and that simulated current (1995) quality (that is the value, had the GAP not taken place). For users within one kilometre this is calculated as Rs 6 million per annum. With pilgrims included, the total becomes Rs 1556 million per annum.

Note that there are two interpretations of the benefits of having the GAP today: (1) the incremental benefits from 1985 to 1995; and (2) the difference between the benefits from the actual current (1995) quality and those from the simulated current (1995) quality (without the GAP). Using the results from the econometric model with quality included as an explanatory variable, the value for users within one kilometre for (1) is Rs 5 million and for (2) is Rs 6 million. The latter is thus around 20 per cent higher than the former, which is the value based on the difference between actual 1985 and 1995 water qualities. When pilgrims are included, the corresponding total values are (1) Rs 1276 million per annum and (2) Rs 1556 million per annum.

The benefits of having the GAP (Phases I and II) may be interpreted as the difference between the value of aggregate benefits from the water quality simulated for 1995 and those from having uniform bathing quality in the river. This is Rs 33 million per annum for users within one kilometre and Rs 6927 million per annum for pilgrims, giving a total of Rs 6960 million per annum.

Note that there are again two ways of interpreting the benefit of having the GAP (Phases I and II): (1) as the difference between the past (1985) value of benefits and those from achieving uniform bathing quality, which is Rs 32 million per annum for users within one kilometre and Rs 6680 million per annum including pilgrims; and (2) as the difference between the simulated (1995) value of benefits and the benefits from having uniform bathing quality water in the river, which is Rs 33 million per annum for users within one kilometre and Rs 6960 million per annum including pilgrims. The latter total value is approximately 4 per cent higher than the former.

COMPARABLE ESTIMATES OF WTP FOR CLEANER WATER QUALITY

There are some willingness-to-pay estimates for improved water quality from valuation studies done in developing countries.[5] These are not always directly comparable, but are presented here to provide a comparative perspective.

URUGUAY

A contingent valuation survey carried out among 1500 randomly sampled households in Montevideo city in Uruguay in 1988–9 sought to measure willingness-to-pay (WTP) for improvements in water quality at beaches near the mouth of a river carrying untreated sewage (McConnell and Ducci, 1989). A project to divert untreated water far enough away from the beaches to eliminate the pollution of the beaches near the mouth of the river would allow swimming and other water (recreation) activities. The household WTP for these direct-user benefits was US$ 14 per annum which, at the 1996 exchange rate of Rs 35 to a US dollar, is Rs 490 per year. This was deemed to be low by local standards. One explanation was that the chosen payment vehicle of a municipal tax was an unpopular option and may have induced strategic bidding by the respondents.

THE PHILIPPINES

A contingent valuation survey of 581 households in Davao City in the Philippines in 1992 sought to measure WTP for improving water quality to swimming (bathing) quality in the river and along beaches close to the city (Choe, Whittington, and Lauria, 1994). No specific plan to clean up the beaches was outlined. Households which used one of the

---

[5]All the studies considered here are reported in a recent draft report to the United Nations Environment Programme, which is listed in the bibliography under Pearce *et al.* (1994).

main beaches in the area, Times Beach, were willing to pay about 30 pesos a month or 360 pesos a year for these direct-user benefits. At 1992 exchange rates of 25.512 pesos per US dollar and Rs 26.412 per US dollar, this amounts to Rs 372.7 per year or Rs 481.61 per year at 1995 prices. Again, this was considered a low value, and a likely explanation was that reducing water pollution was not a high priority for the residents of the area.

CHINA

A valuation of the benefits of improving water quality in Lake Tai in the Wuxi province of China to swimming quality was carried out in 1991–2 among residents of the neighbouring community (Abelson, 1996). Households living nearby were willing to pay US $ 15 per annum for these direct-user benefits which, at the current exchange rate of Rs 35 to 1 US dollar, is about Rs 525 per year.

EGYPT

Direct-user benefits from improvements in water quality in Lake Timsah in Ismailia, Egypt were estimated in 1987 (Luken, 1987; quoted in Winpenny, 1991). Using the travel cost approach, the consumer surplus measure of annual direct-user benefits from the clean-up was £ (Egyptian) 16.35 to 21.40. At 1987 exchange rates of 1.273 Egyptian pounds per US dollar and 12.968 Rupees per US dollar this amounts to a range from Rs 166.51 to Rs 217.99 or from Rs 346.72 to Rs 453.91 at 1995 prices.

COLOMBIA

The direct user-benefits to households from cleaning up a feeder river into Rio Bogota which skirts Bogota, the capital city of Colombia, were estimated in a study done at the end of the 1980s (ODA, 1990). The total household willingness-to-pay for improved health and amenity benefits, as well as for the convenience of being connected to a modern sewerage system, worked out to US$ 41 million per annum. At the 1996 exchange rate of Rs 35 to a US dollar, this is a *total* of Rs 1435 million. However, this was not an estimation based on household responses, but 5 per cent of average household income in the area was taken to be the *assumed* willingness to pay (Winpenny, 1991: p. 196).

For purposes of comparison, these estimates are presented in Table 6.11, along with the relevant estimates from the present contingent valuation survey of direct-user benefits.

TABLE 6.11

Comparable Estimates of WTP for Direct-User Benefits from
Improved Water Quality

Country	Year	Household WTP (Rupees per household per annum)	Total WTP (Rupees million per annum)
Uruguay	1988–9	490	
The Philippines	1992	481	
China	1991–2	525	
Egypt	1987	347–454	
Colombia	Late 1980s		1435*
India	1996	488–510	16–16.5

*based on assumed not estimated willingness-to-pay.

CONCLUSIONS

This study is an attempt to estimate the user benefits of cleaning a major water resource like the Ganga, which has a very high economic significance in India. The Contingent Valuation Method is used with a carefully designed questionnaire for surveying a sample of urban households living near the banks of the river in the cities of Calcutta, Varanasi, and Allahabad. It attempts to measure only user benefits accruing to urban populations living near the Ganga as it flows through the three states of Uttar Pradesh, Bihar, and West Bengal. There can obviously be substantial user benefits accruing to rural households living near the river in its entire course of 2525 kilometres, but this study does not measure those benefits.

As in the case of the valuation by urban households of non-user benefits from levels and changes in water quality in the Ganga described in the previous chapter, urban households deriving direct-user benefits have also shown consistent preferences for the quality of water in the Ganga. The estimated willingness-to-pay (WTP) function relating household WTP to the quality of water in the river and socio-economic variables representing household characteristics shows that household WTP increases with the quality of the river, as well as with the income size and environmental awareness of households. The education level of the respondents is once again a proxy for their environmental awareness. And the fact that the sample consisted of households from the literate urban population in the country no doubt played a major role in generating the good response rate of the survey.

As in the case of the contingent valuation survey of non-user benefits reported in the previous chapter the estimation results from the econometric model with water quality included as an explanatory variable proved a good measure of fit to the data with all the variables having the correct sign (denoting consistent preferences) and highly statistically significant coefficients. Calculations of household and aggregate willingness-to-pay for different levels and for changes in river water quality are based on the results from this model.

The main results are as follows. The estimate of willingness-to-pay (WTP) for user benefits from bathing quality water is Rs 581.59 per household per annum, which aggregates to Rs 19 million per annum for the 37213 household in the target population of literate urban residents in the major cities in the three states of Uttar Pradesh, Bihar, and West Bengal who live within 0.5 kilometres of the river.

Willingness-to-pay for benefits from current (1995) water quality in the Ganga is Rs 167.23 per household per annum, which totals to Rs 5.5 million per annum.

Similarly, household WTP for the direct-user benefits from the level of water quality existing in 1985 (that is before the GAP started) is Rs 93.28 per annum, which sums to Rs 3.1 million per annum.

Reviewers of the study noted that the WTP figures from the user and non-user surveys are similar (around Rs 500 for bathing water quality and around Rs 150 for current water quality). One might expect higher values for users than for non-users, and one possible explanation offered was that the average incomes of literate households along the Ganga are lower than those of households living elsewhere (since UP and Bihar are poor states). Thus, users may be willing to pay a higher percentage of incomes, even if the absolute amounts are similar to other places. The average per capita incomes of user households is Rs 27,771 compared to Rs 36,467 for non-users, which is consistent with the above observation. Furthermore, it has been suggested that there may have been questionnaire design problems that could have resulted in the similar figures. In this context, it is interesting to note that the values obtained for user benefits are not dissimilar to other studies. For the Gomti, for example, a recent DFID study reported a WTP of between Rs 100 and Rs 200 in user benefits for a cleaner river. This suggests that the bias referred to above is likely to be limited.

As in the case of the non-user analysis, the econometric model with water quality makes a further calculation possible. Using an Index of River Water Quality developed based on the data from the National Rivers Conservation Directorate, Ministry of Environment and Forests, and simulations on the Water Quality Model done by the Industrial Toxicology Research Centre, Lucknow (see previous chapter), the direct-user benefits of the water quality that would have existed in 1995 *in the absence of the GAP* can be worked out. In this simulated scenario, in which there is no GAP, the estimated household WTP for 1995 quality of the river is Rs 71.12 per annum, which aggregates to Rs 2.3 million per annum.

Using these mean WTP values for user benefits, the benefit of having the GAP today can be estimated in two ways: (1) the incremental benefits from 1985 to 1995; and (2) the difference between the benefits from the actual current (1995) quality and those from the simulated current (1995) quality (without the GAP). The former is Rs 2.4 million per annum and the latter is Rs 3.1 million per annum, the latter being about 30 per cent higher than the former.

Similarly the benefit of having the GAP (Phases I and II) can be interpreted in two ways: (1) as the difference between the past (1985) value of benefits and those from achieving uniform bathing quality; and (2) as the difference between the simulated (1995) value of benefits and the benefits from having uniform bathing quality water in the river. The former is Rs 16 million per annum, while the latter is about 4 per cent higher at Rs 16.7 million per annum.

A comparison of these WTP estimates with those from other developing countries, of direct-user benefits of *improvements* in water quality levels, shows that the values obtained in this contingent valuation are broadly comparable.

7

The Health Impacts of the Ganga Action Plan

INTRODUCTION

This chapter deals with the health effects of the GAP. Improvements in health are a serious factor behind the concern with water quality; hence the health effects of a major programme such as the GAP are important in any cost–benefit analysis of its effects. Unfortunately, the quantification of these benefits is difficult, especially for programmes such as GAP which deal with improving river water quality in a general sense and do not focus on drinking water or sanitation.

The next section discusses the evidence of the benefits arising from improvements in water quality on health from other studies in the sub-continent, and in other developing countries. It also reports some estimates of the monetary value attached to those impacts. The following sections report are based on the results of a study by Professor S. Nath an his colleagues at AIIH & PH on some monetary estimates of the putative health benefits of the GAP and the analysis of data collected for measuring the health effects of the GAP. These data are analysed for differences in health effects between areas that have been affected by the GAP and those that have not, and some preliminary conclusions are drawn. These are presented and critically reviewed.

THE HEALTH BENEFITS OF WATER QUALITY IMPROVEMENT

GENERAL PRINCIPLES

Health damage studies should proceed first by establishing average levels of ambient concentration of each pollutant. The next stage is to relate those concentrations to health effects, or 'end-points', such as premature mortality and morbidity. For this, ideally, dose-response functions (DRFs) are required. These are taken from the epidemiological literature. DRFs may have to be estimated in two ways: first by following a given cohort of people through time and recording their health status. This is then related to a time series of pollution concentration data. The other approach is cross-sectional, for example correlating health data in different areas within cities, or even across cities, with factors that are likely to explain variations in that health status. Both of these methods make the—not necessarily valid—assumption that ambient concentrations reflect true exposure. Consensus DRFs are possible if there are sufficient reliable studies that provide similar quantitative estimates of the impact of a change in ambient concentrations on the health of the population at risk.

In the health literature, the DRFs are much better developed for air pollution than they are for water pollution. The difficulties with the DRFs for water are clear when we consider the studies that have been undertaken hitherto in India on this subject. We found two studies that were relevant to the subject of this study—one a 'macro' level study and the other a detailed 'micro' level study.

THE BRANDON–HOMMANN AND OTHER WORLD BANK STUDIES

Brandon and Hommann (1995) made a rough estimate of the monetary damages from water pollution in India. Before discussing their results, it is instructive to look at the underlying studies on which they relied for the DRFs.

A large number of studies have shown that improved water quality does improve human health. Barter and Hommann cite the WHO survey of 144 studies (Esrey *et al.*, 1991), which showed that improved water supply and sanitation produced a median reduction in the order of 25 per cent for morbidity and 65 per cent for mortality. The statement implies that improvements for the 'average quality' to be found in developing countries, to water that is pristine, would result in the above health benefits, although that is not altogether clear. Since the actual quality level will vary from country to country and the improvements achieved will also differ according to the programme, it is extremely difficult to use generic studies of this kind to evaluate a particular programme such as GAP.[1] Furthermore, also implicit in the analysis is

[1] A meta analysis takes a large number of individual studies and uses them to see how the effects of changes in ambient concentrations vary with local conditions and other factors.

the assumption that there are clear thresholds: water above a certain quality is safe and below that quality it is not. This assumption is also controversial: the health impacts of poor quality water are not independent of 'how poor' it is; nor is there a distinct point above which further quality improvements will yield no benefits (at least in the range of quality involved in this study).

Brandon and Hommann use the above estimates to derive a figure for the number of disability adjusted life years (DALYs) lost due to poor water quality, sanitation, and hygiene. They conclude that DALYs of the order of 30.5 million are lost each year. The most important specific diseases contributing to this loss are diarrhoea, trachoma, intestinal worms, hepatitis, and a broader 'tropical cluster'[2] of diseases.

The detailed estimate of water-borne diseases in India, on which the above estimate of 30.5 million is based, is given in a data set assembled by the World Bank/World Health Organization (WHO) (World Bank, 1995) which shows that about 21 per cent of all communicable diseases in India (11.5 per cent of all diseases) are water related. Tables 7.1 and 7.2 give the estimated DALYs for the different diseases.

TABLE 7.1

Incidence of Water-borne Diseases by Cause in India, 1990

(hundreds of thousands of DALYs lost)

Disease type	Female	Male	Total
Diarrhoeal diseases	143.9	136.4	280.3
Acute watery	78.9	75.0	153.9
Persistent	42.6	40.2	82.8
Dysentery	22.4	21.3	43.7
Hepatitis	1.7	1.4	3.1
Tropical cluster	7.5	11.3	18.8
Schistosomiasis	0.9	1.7	2.6
Leishmaniasis	5.0	6.8	11.8
Lymphatic filiarisis	1.6	2.8	4.4
Trachoma	2.0	1.1	3.1
Total	155.1	150.2	305.3

[2]This contains the following diseases: Trypanosomiasis, Chaga's disease, schistosomiasis*, leishmaniasis*, lymphatic filariasis*, onchocerciasis. Only those marked with an asterisk are relevant to India.

There are a number of problems in making use of the above data. The main one is that the link between the water-borne diseases and a quantitative measure of water quality is not available. The GAP and other water projects make improvements in water quality, which necessarily result in levels that are 'safe', however that may be defined. Hence, without some way of linking the quality measures to the disease levels we cannot estimate the health impacts.

TABLE 7.2

Incidence of all Diseases by Consequence, Sex, and Age in India, 1990

(millions of DALYs lost)

| Age Group | *As a result of premature death* | | | *As a result of disability* | | |
	Males	Females	Total	Males	Females	Total
0–4	53.0	55.1	108.1	14.2	14.9	29.1
5–14	9.4	10.8	20.2	6.4	5.7	12.1
15–44	18.9	17.7	36.6	12.0	17.2	29.2
45–59	9.8	7.3	17.2	6.7	5.0	11.7
60+	9.7	8.7	18.4	5.2	4.6	9.8
Total	*100.8*	*99.7*	*200.6*	*44.5*	*47.4*	*91.9*

A second, equally serious problem is that water quality is relevant to humans through a number of pathways and the above data do not indicate which pathways are the most important or, more generally, they do not indicate the relative importance of the different pathways. It is known that water pollution has three major sources which can infect both surface and ground water: domestic wastewater, industrial wastewater, and agricultural run-off. Water pollution from domestic and human wastewater is the most problematic and the cause of much water-borne disease transmission. Major cities dispose some of their untreated sewage into irrigation streams used to irrigate crops, some of which are eaten raw. Sewage and wastewater is also channelled into rivers and streams without consideration of the rivers' assimilative capacity. The other sources of water pollution are industry and agriculture. The major water polluting industries are chemicals, textiles, pharmaceuticals, cement, electrical and electronic equipment, glass and ceramics, pulp and paper board, leather tanning, food processing, and petroleum refining. Indiscriminate use of agricultural chemicals has also resulted in the contamination of ground and surface water. The health impacts of industrial and agricultural pollutants cannot be separated easily from the overall health impacts. In terms of water quality improvements, it is most likely that improved sanitation and sewage facilities will have a bigger impact on health than improvements in river quality, but the latter may not be insignificant and we have no way of knowing from the above data how big the effect will be.

THE VERMA–SRIVASTAVA STUDY

The Verma–Srivastava (1990) study is the only micro-scale study to date, containing data which might be useful for the cost–benefit analysis of cleaning the Ganges. The authors state that they were unable to trace any previous research which made estimates of the costs of the water related diseases anywhere else in India. They undertook an investigation into water related diseases in the district of Jhansi, one of the poorest rural areas of Uttar Pradesh. The methodology and main results of this study are outlined below.

Data collected was for a 20 per cent random sample stratified by occupation type (agriculture, labour, business, service, artisan, household, child/student, other). It covered sex and age; general socio-economic and demographic characteristics; village population, immigration, and emigration; annual income; and health.

Health was assessed by weekly visits to record information about any illness in the preceding week by interview with the senior household member using a standardized questionnaire recording symptoms. Periods lost from work due to illness were recorded in days, with days of productivity lost from illness differentiated from days of illness. The illnesses which were considered in the study were: enteric fever, diarrhoea, dysentery, gastroenteritis/cholera, infective hepatitis, trachoma, conjunctivitis, and scabies. There are many other water-borne diseases, but those chosen were included in view of their appropriateness in health impact studies of water supply programmes, after reviewing a series of studies on the subject.

An illness was considered to have ended when the symptoms had not been recorded on two consecutive days. Symptoms were grouped as an illness after review of the household charts by the field investigators, with assistance from the medical epidemiologist where necessary. The main problem here was to ensure that the correct diagnosis of water related disease was made from the symptoms. The main results of estimated health effects are summarized in Table 7.3.

TABLE 7.3

Days Lost due to Illness from Water Related Diseases in Village People in Jhansi District, Uttar Pradesh, 1981–2

Disease	Number of days lost per annum		Number of days lost (per 100 people)	
	1981	1982	1981	1982
Enteric fever	25,500	29,466	423	479
Acute diarrhoeal disease	23,424	23,088	389	375
Infective hepatitis	741	1216	12	20
Conjunctivitis	37,692	16,191	626	263
Scabies	4800	25,092	80	408
All diseases	*92,157*	*95,053*	*1529*	*1544*

It should be noted that small-scale village level surveys are very sensitive to local epidemics of particular diseases. This is reflected in the much higher incidence of conjunctivitis in 1981 and of scabies in 1982. If allowance is made for these epidemics, the rank order of importance of particular diseases would be: enteric fever, followed by acute diarrhoeal diseases, conjunctivitis, scabies, and then infective hepatitis.

Unfortunately, we cannot make any direct use of this study as there are no data reported on water quality. Nevertheless, it is useful as a check against the kinds of estimates that are provided in the next section.

VALUING HEALTH BENEFITS OF WATER QUALITY IMPROVEMENTS

Once the health impacts have been identified, it is necessary to value them in monetary terms if a full cost benefit analysis is to be completed. The methods of valuation are discussed briefly in the chapter on methodology.

THE BRANDON–HOMMANN STUDY

In order to quantify the achievable reductions in the 30.5 million DALYs due to the provision of clean water and sanitation, a methodology developed at the World Bank was applied.[3] This methodology involves several steps:

(a) Since available data on water related health impacts in India come in the form of DALYs, it does not allow separate analysis of mortality and morbidity. Therefore, potential reductions achievable in mortality (65 per cent) and morbidity (25 per cent) are combined into one reduction factor to be applied to the current level of DALYs for water related communicable diseases. By using a weighted average of the overall mortality and morbidity components of these DALYs, it is estimated that an overall reduction of 52.4 per cent in DALYs is possible with the provision of safe water and sanitation. This implies that 47.6 per cent of the current level of water related DALYs represent a 'floor', below which overall public health will not improve unless current hygiene related practices improve. All the scenarios explored below assume that hygiene improves marginally with access to water and sanitation (see point 'd' below).

(b) A relationship is postulated between the incidence of DALYs among populations with and without safe water and sanitation. This may be expressed as

$$P * R_w + (1 - P) * R_{w0} = R_t, \tag{7.1}$$

where P is the proportion of population with adequate services in the form of safe water and sanitation; R_w is the water related DALYs lost per thousand among those with clean water and sanitation; R_{w0} is the water related DALYs lost per thousand among those without clean water and sanitation; and R_t is the average water related DALYs lost per thousand overall.

Since it is known that the total water related DALYs in India are 30.51 million, and the total 1990 population is 850 million, this means that

$$R_t = 30.51/850 * 1000 \quad \text{or} \quad 35.9. \tag{7.2}$$

(c) As of 1990, 73 per cent of the population in India had safe water, but only 14 per cent had adequate sanitation (World Development Report, 1994). Since these two numbers are very different, there is no single value of P that captures the hygienic effect of having safe water and sanitation. Depending on the relative importance of water versus

[3]'Environmental Strategy for the Middle East and North Africa', World Bank, grey cover, 15 February 1995, especially Annex 5.

sanitation in reducing mortality and morbidity, P can assume any value in the range of

$$x * 0.73 + (1 - x) * 0.14, \qquad (7.3)$$

where x is between 0 and 1.

In the estimates below, a value of $x = 0.9$ (and hence, $P = 0.67$) assumes that water is relatively more important for disease reduction, whilst $x = 0.5$ (and hence, $P = 0.44$) assumes that both water and sanitation are equally important.

(d) Another issue relates to the role of hygiene levels in determining the relation between R_w and R_{w0}. Evidence worldwide associates both higher hygiene levels and better access to clean water and sanitation with higher literacy and income levels. This implies that hygiene levels are lower among the population with lower safe water and sanitation provision rates, since these rates and literacy are lower in rural areas. To address the positive correlation between clean water and sanitation and hygiene, a hygiene factor h is introduced

$$R_w = R_{w0} * 0.476 * h, \qquad (7.4)$$

where $h \leq 1$, and the parameter 0.476 comes from step (a). $h = 1$ implies that the poor are equally hygienic as those with better services, which is unrealistic. Values of h in the range of 0.6–0.8 may be more realistic, and are used here to illustrate the likely range of health improvements likely to emanate from improved hygiene.

The value of the water related mortality and morbidity has been estimated using the human capital approach. The statistical value of one DALY is taken as equal to the annual average productivity of Indian workers (since one DALY implies one year in which a worker cannot work due to either sickness or premature death). This value is often adjusted for the age of the DALY, where the value rises from birth to young adulthood, and declines after 50 years of age. Since the incidence of water related DALYs falls heavily on children under the age of 5 years, the lower estimate of the value of one DALY is based on a weighted average of the average statistical value of life weighted by the age distribution of water related DALYs. The low value assumed is $215 per DALY. A medium estimate, $330 per DALY, assumes a constant statistical value of life across all ages.

As was emphasized in the chapter on methodology, Chapter 3, we do not favour a valuation system for health impacts that is based on productivity. A better method is based on willingness-to-pay (WTP), from which the value will almost always be higher than value derived from the productivity approach. Such a higher estimate has been derived based on the value of a statistical life, which was discussed in the chapter on methodology and which can also be seen as the value attached to the loss of a number of years of life, depending on age. It is based on a WTP approach and methods have been developed for converting it to a value of life years lost. One estimate for a DALY in India is $570 per DALY, based on a ratio of the US Environmental Protection Agency or US–EPA annualized value of life converted to the Indian context by applying the ratio of national per capita incomes, weighted by the distribution of DALYs across age brackets (as for the lowest estimate). Applying these values to the number of DALYs lost gives a total estimate for the country of $3.1 to $8.3 billion per year, depending on the assumed value of one DALY ($215–570). If services are extended to only half of those in need, assuming average values for the parameters, the health savings would be between $1.5 billion and $4.2 billion per year.

This kind of 'macro' estimate of the benefits of water quality improvement is useful as a general check in special studies of the kind carried out here, but it is not a substitute for a micro-level estimation of the costs. Nor does it provide the parameters that we can use in the micro-level study.

THE VERMA–SRIVASTAVA STUDY

In the Verma–Srivastava study, the costs due to water-borne illness were estimated by multiplying daily income by the total number of days lost due to water related illness in a year. These losses were added to the cost of treatment (that is expenditures such as doctor's fee, medication, X-ray, travel to clinic/hospital). In this method there are difficulties in incorporating full cost of treatment when medical care is provided free by the state.

The estimates obtained for water-borne diseases are given in Table 7.4.

An annual cost of around Rs 248 per person at 1981 prices converts to an annual cost of water-borne diseases of about Rs 915 at 1995 prices.

As with the macro estimates of water-borne disease cost, this study does not provide us with any directly transferable parameters for the GAP project. The main reason is the lack of information on water quality and the link between water quality and the level of disease.

TABLE 7.4
Estimated Total Cost of Illness to Village People in Jhansi District,
Uttar Pradesh, Due to Water Related Diseases in 1981–2

(Rupees in 1981 prices)

Disease	*Estimated annual loss of income (per 100 people)*		*Estimated expenditure (per 100 people) on treatment*		*Estimated total cost (per 100 people)*	
	1981	*1982*	*1981*	*1982*	*1981*	*1982*
Enteric fever	5663	6657	1690	1955	7353	8622
Acute diarrhoeal diseases	4846	4704	486	487	5333	5191
Infective hepatitis	163	239	48	83	211	323
Conjunctivitis	6573	3072	491	217	7364	3289
Scabies	1076	5287	763	2115	1839	7402
All diseases	*18,622*	*19,969*	*3478*	*4858*	*22,100*	*24,827*

Source: Verma and Srivastava (1990).

ESTIMATES OF THE HEALTH BENEFITS OF THE GAP

This section looks at the health benefits resulting from improvements in water quality achieved by the Ganga Action Plan. In particular, it draws on the study conducted by the All India Institute of Hygiene and Public Health (AIIH&PH, 1997) which presents 'before' and 'after' the GAP data on the health of populations living along the Ganga river. Valuation estimates for the health benefits of improved water quality from the GAP are calculated based on:

1. Improvements to users' income due to reductions in working days lost.
2. Data from a contingent valuation survey measuring willingness-to-pay for improved water quality.
3. Cost of treatment of water for public supply without implementation of GAP.

AIIH&PH STUDY

The AIIH&PH study selected areas of six towns on the Ganges to conduct a health survey and a contingent valuation survey. For the towns of Hardwar, Kanpur, Chandannagar, and Titagarh, corresponding studies were conducted among a group of controls in the same towns with similar characteristics as the study areas, except that they had not been affected by the GAP, selected for comparative evaluation with the study areas. For the towns of Patna and Nabadwip no such control studies were undertaken.

Morbidity data were collected for a six month period while mortality rates are given for 12 month periods. As this was a cross-sectional study, the seasonal variation in the prevalence and impact of water-borne diseases was not assessed. Since the study was undertaken during the monsoon period in Nabadwip and Titagarh, and during the winter for the other four towns, the results in these two towns may not be directly comparable with rates given for the other towns, for example, in the case of diarrhoea.

Hardwar

The death rates per thousand of population were 61 in the study area and 73.4 in the control area—both very high figures. In general, the proportional death rates due to water-borne diseases are higher in the control than in the study area, for example, the rates for diarrhoea are 8.20 per cent and 3.96 per cent respectively. Similarly the 6 month water-borne disease prevalence percentages are generally less in the study area than the control area, for example, for diarrhoea the rates are 0.65 per cent and 0.71 per cent respectively. However, for helminthic infection and poliomyelitis the rates are higher in the study area.

The number of working days lost due to sickness is higher in the control area. Average loss of working days in the control area is 2.3 per household per six month period, whereas in the study area it is 1.4. Expenditure on medicines and doctors is also greater in the control area, at Rs 155 per household per six month period compared to Rs 111 for the study area.

Chandannagar

The death rates per thousand of population were similar in both areas (24.5 in the study area and 26.4 in the control area). The proportional mortality rate due to water-borne diseases is low at 6.06 per cent. However, this was still more than the control area, which had a mortality rate of 0.00 per cent. The prevalence of water-borne diseases was generally greater in the study area than in the control area, with diarrhoea and helminthic infection rates of 5.58 per cent and 6.47 per cent respectively, compared to 2.76 per cent and 2.76 per cent in the control area. The average number of working days lost per household per six months is greater in the study area (0.75 days) compared to 0.5 days in the control area. Expenditure on medicines and doctors in the study area (Rs 140) was slightly higher than the control area (Rs 111) per household per six months.

Kanpur

The crude death rate in the study area was 27.6, compared to 20.6 in the study area. The mortality breakdown shows that 6.25 per cent of deaths were due to diarrhoeal diseases in the study area, compared to none in the control area. The prevalence of water-borne diseases was also greater in the study area for all the diseases. As expected, the number of working days lost was much higher in the study area (2.8 per household per six months), against 0.9 for the control area. The same applies to expenditure on medicines which runs to Rs 214 per household per six months in the study area and to Rs 85 in the control area.

Patna

There was no control area for Patna against which the study area can be compared. Some general comparisons with other control areas and other study areas can, however, be made. The crude death rate here was relatively low, at 18.7 per thousand . Deaths from water-borne diseases were not exceptional compared to other control areas. The six month morbidity prevalence due to water-borne diseases was relatively low for diarrhoea and skin diseases and relatively high for helmenthic infection. Expenditure on medicines and doctors runs to around Rs 100 per household per six months.

Titagarh

The crude death rates in Titagarh were similar in the control and study areas (9.8 per thousand against 10.6). The percentage of deaths from water-borne diseases was quite high in the study area compared to the control area (27 per cent against none for the control area). The prevalence of water-borne diseases was also uniformly higher in the study area. Loss of working days was, however, similar in both areas, at about 0.45 days per household per six months. Average household expenditures on medicines and doctors were remarkably different, at Rs 147 in the control area and Rs 19 in the study area. The latter figure is so out of line with other data that it is somewhat suspect.

Nabadwip

There was no control area in Nabadwip. The crude death rate was high, at 40.4 per thousand. The proportional mortality from water-borne diseases was also very high (at around 40 per cent of all deaths, it is the highest among all the samples). The prevalence of water-borne illnesses was also high, especially for diarrhoea, helmenthic infection, and skin diseases. Loss of working days, however, was very low (3 per household per six months), but expenditure on medicines and doctors was the highest among all the samples (Rs 256 per household per six months).

Statistical tests on differences in means

The data collected in the above samples have been tested for statistical significance of difference in means for the localities where data are available for both, a study and a control area. The results are given in Table 7.5. The table reports on whether the mean rate in the study area is statistically significantly different from the mean rate in the control

TABLE 7.5
Statistical Difference Between Morbidity Rates in Control
and Study Areas

Area	Diar-rhoea	Enteric fever	Infective hepatitis	Helminthic infection	Polio-myelitis	Skin disease	Other
Haridwar	NO	YES +	NO	NO	NO	YES +	NO
Kanpur	YES –	NO	NO	NO	NO	YES –	YES –
Chandannagar	YES –	NO	NO	YES–	NO	YES –	YES –
Titagarh	YES –	NO	NO	YES–	NO	YES –	YES

Source: See text.

area. A negative sign indicates that the control area is statistically significantly lower than the study area, while a positive sign indicates that the control area is statistically significantly higher than the study area.

Table 7.5 shows that approximately half of the morbidity rates (13 out of 28) were statistically significantly different from each other (control against study area). Of these, only 2 have a mean rate of morbidity in the study area that is lower than the control area. In 11 cases the mean rate in the study area is higher than the control area. This makes one inclined to conclude that, in general, the morbidity rates in areas affected by the GAP are still higher than in the control areas and there is no evidence from this data that morbidity from water-borne diseases has been reduced as a result of the GAP.

BENEFITS OF THE GAP DUE TO REDUCTIONS IN WORKING DAYS LOST THROUGH SICKNESS

Pre-study period morbidity data for Varanasi were compared to post-project data for the study areas of Patna, Kanpur, and Hardwar in order to get an indication of the impact on health of improvements in water quality in the river Ganga. The results are summarized in Table 7.6 and give total water related morbidity rates of 1.93 per cent for Hardwar, 14.46 per cent for Patna, and 33.68 per cent for Kanpur, compared to a pre-project figure of 38 per cent for Varanasi. The post-project rates of water-borne diseases are all lower than the pre-project rate, but there are considerable difficulties in interpreting the data. First, the pre-project area is not the same as the post-project areas, and although the post-project study areas were selected so that they are similar to the pre-project area, other factors may be responsible for the differences in morbidity rates. Second, other factors have intervened over time which may be responsible for some of the difference (for example increases in income or provision of sanitation).

Similar comparisons were made for the number of working days lost due to sickness between pre-project data for Varanasi and post-project data for Hardwar, Patna, Kanpur, Titaghar, Chandannagar, and Nabadwip. The results are summarized in Table 7.7 (Hardwar, Patna, and Kanpur) and show post-project figures for the study areas significantly below the pre-project figures for Varanasi. Average household loss of working days was 4.1 per six months in Varanasi pre-project, and was 1.1 in Patna, 2.5 in Kanpur, and 1.3 in Hardwar, post-project.

TABLE 7.6

Comparative Morbidity in Households (six month study periods):
Varanasi vs Patna, Kanpur, and Hardwar

Disease	Pre-project	Post-project		
	Varanasi (%)	Patna (%)	Kanpur (%)	Hardwar (%)
Diarrhoea	22.1	3.82	11.44	0.65
Enteric fever	2.9	6.31	6.17	0.53
Infective hepatitis	1.6	3.82	1.54	0.09
Helminthic infection	5.5	0.35	7.33	0.43
Poliomyelitis	0.2	0.08	–	0.14
Skin disease	5.7	0.08	7.2	0.09
Total	*38.0*	*14.46*	*33.68*	*1.93*

TABLE 7.7

Comparative Loss of Working Days Due to Sickness, Pre- and
Post-Project (percentage of households affected during six month
study periods): Varanasi vs Patna, Kanpur, and Hardwar

Days	Pre-project	Post-project		
	Varanasi (%)	Patna (%)	Kanpur (%)	Hardwar (%)
Up to 2	6.95	0.6	4.66	–
3 to 4	13.34	0.9	11.33	2.54
5 to 6	7.26	1.8	4.66	1.69
7 to 8	3.39	0.6	2.00	0.28
9 to 10	7.34	0.3	5.33	0.28
11 to 15	5.63	3.0	3.33	1.41
16 to 20	2.31	–	1.33	0.28
21 to 25	–	–	–	0.28
26+	3.1	1.5	2.00	2.54
Average	*4.1*	*1.1*	*2.5*	*1.3*

The same caveats apply, however, to this comparison as to the comparison of the morbidity figures.

For the areas of Nabadwip, Titagarh, and Chandannagar, a comparison of post-project data was made with pre-project data from Nabadwip. The relative morbidity rates are given in Table 7.8 and the corresponding loss of working days in Table 7.9. In this case, the morbidity rates are

TABLE 7.8

Comparative Morbidity in Households (six month study periods):
Nabadwip vs Titagarh, Chandannagar, and Nabadwip

	Pre-project	Post-project		
Disease	*Nabadwip* *(%)*	*Titagarh* *(%)*	*Chandannagar* *(%)*	*Nabadwip* *(%)*
Diarrhoea	10.89	9.42	5.58	11.39
Enteric fever	1.26	1.53	2.08	1.83
Infective hepatitis	0.42	1.4	0.82	1.1
Helminthic infection	14.87	10.71	6.71	12.86
Poliomyelitis	0.09	0.59	–	0.36
Skin disease	6.17	9.2	6.25	7.35
Average	*33.7*	*32.85*	*21.44*	*34.89*

TABLE 7.9

Comparative Loss of Working Days Due to Sickness, Pre- and Post-Project (percentage of households affected during six month study periods): Nabadwip vs Titagarh, Chandannagar, and Nabadwip

	Pre-project	Post-project		
Days	*Nabadwip* *(%)*	*Titagarh* *(%)*	*Chandannagar* *(%)*	*Nabadwip* *(%)*
Up to 2	0.79	4.03	2.02	1.58
3 to 4	4.38	2.41	2.53	–
5 to 6	11.95	–	1.77	7.93
7 to 8	11.55	14.03	1.26	6.34
9 to 10	9.56	–	0.51	4.76
11 to 15	4.78	–	1.01	3.17
16 to 20	0.79	–	–	3.17
21 to 25	4.52	–	–	3.17
26+	0.29	–	0.76	–
Average	*4.5*	*1.2*	*0.7*	*3.1*

slightly lower post-project for Titagarh and Chandannagar, but higher for Nabadwip. This is attributed to the fact that for Nabadwip, the water quality around the bathing ghats has not improved and the environmental sanitation situation remains poor. Average working days lost, however, were 4.5 per household per six months for Nabadwip (pre-project) and were 1.2 for Titagarh, 0.7 for Chandannagar, and 3.1 for Nabadwip (post-project).

The yearly benefit of improvements in water quality of the river Ganga due to the GAP was then calculated in terms of increases in household incomes for each of the six study areas (where average number of members in a household is assumed to be five). This is based on the difference between the number of working days lost due to sickness for the pre-project site and the number lost in post-project study areas. These figures for decreases in the number of household days lost are multiplied by the average daily income for each study area calculated using data from the users' benefit surveys, to arrive at figures for yearly benefits for each study area. The results were as shown in Table 7.10.[4]

The total number of users in the six study areas is 473,550 and the total benefits estimated are Rs 34.84 million. Hence, the average annual benefit of improvements to water quality based on reduction in working days lost through sickness is calculated as Rs 73.57 million per million users.

TABLE 7.10

Health Benefits from Reduced Loss of Working Days

Town	Average No. of working days saved yearly per family	Ragular users of Ganga in town (No. of families)	Individual daily income (Rs)	Total value (Rs mn)
Hardwar	6.09	41,300	64	16.00
Kanpur	3.42	15,560	35	1.86
Patna	6.58	22,480	88	12.94
Chandannagar	6.44	5,690	38	1.37
Nabadwip	7.37	4,760	65	2.28
Titagarh	2.61	4,920	31	0.39

Source: AIIH&PH (1997).

[4]The data provided by AIIH&PH was for household incomes. Since the loss of days is for person days, the value per person can be estimated by dividing the household income by the size of the household. It is assumed that individual incomes are half household incomes—i.e. there are two full time earners per household.

STUDY OF SEWAGE FARM WORKERS

A further comparative study concentrated on the health conditions of the sewage farm workers of Titagarh before and after water quality improvement schemes. The results of a stool examination indicated that worm infestation was reduced in workers in the post-project period due to optimal running of the sewage treatment plant. Protozoal infestation was also found to be reduced. These results were used as a basis for a notional estimate of a reduction of four days per month in working days lost per sewage farm worker due to these improvements. Therefore the total monetary benefit was calculated as Rs 480,000 per year per 100 sewage farm workers.

BENEFIT OF WATER QUALITY IMPROVEMENT MEASURED BY SAVINGS IN WATER TREATMENT

The AIIH&PH study also looked at the likely cost of the treatment of water for public supply, had water quality not been improved by the implementation of the GAP. It is likely that without the GAP, the present system of water treatment followed after the GAP would be inadequate to provide water for drinking purposes. Reaching this standard would require the upgrading of treatment technology by at least introducing activated carbon absorption techniques for the removal of toxic chemicals and heavy metals. The additional cost of this technology was calculated for the six towns in the study based on the assumption of the cost of activated carbon absorption techniques at Rs 3,000 per million litres. The results are given in Table 7.11 and show a total benefit per year from implementation of the GAP of Rs 731.29 million for the six towns, or Rs 164.25 per million people. This indicates that in Calcutta, where water supply from the Ganga is 900 MLD (million litres per day), the yearly benefit of improved water quality through the GAP is Rs 985.5 million.

Although the above estimates are useful information, they cannot be included in the cost–benefit analysis as such. The reason is that the study is measuring the improvements in health and other factors resulting from the improvements in water quality. If one then adds the cost of an alternative method of improvement as a benefit, one is double counting. If there had been no GAP, the health and other benefits would not have materialized. Of course, if water treatment costs fall compared to what they were before the GAP, those savings are correctly to be added to GAP benefits. But the above cost savings are relative to a technique that is an improvement in water treatment which achieves the goals of GAP[5].

TABLE 7.11

Benefit of GAP Water Quality Improvements as Measured by
Savings in Water Treatment Expenditure

Town	Population	Water supply* (MLD)	Benefit per year (Rs mn)
Tigagarh	114,085	17.11	18.74
Chandannagar	120,378	18.06	19.78
Nabadwip	104,533	15.68	17.17
Patna	13,76,701	206.51	226.13
Kanpur	20,37,333	305.6	334.63
Hardwar	699,230	104.88	144.84
Total			*731.29*

CONCLUSIONS

This chapter has looked at the health benefits of improving water quality. Previous studies such as those by the World Bank and Verma–Srivastava have come up with estimates of some general benefits, but they are not attributable to particular kinds of water quality improvement; in particular they do not focus on river water quality. Moreover, they do not establish a quantitative link between water quality and health benefits. Nevertheless, their values are instructive. The World Bank study estimates and in particular, the costs of water-borne diseases estimated by Verma and Srivastava are around Rs 915 per household per annum. In this study the AIIH&PH study estimated the benefits of an improvement in water quality as a result of the GAP (which does not result in completely satisfactory water quality), of Rs 73.57 per river user per annum. For a household of 5 members, this would amount to

[5]A reviewer of the draft report has noted that, for the seven cities along the Ganga the annual cost 'savings' are higher than the total benefits calculated in this study. This implies that, in a situation where water supplies have to come from the river, it is likely to be more cost-effective to invest in water treatment plants to clean up polluted river water. There would be merit in examining this issue further.

Rs 367.8, which is less than the estimate by Verma and Srivastava; but in general the two figures are not completely out of line.

In general, the AIIH&PH study is not altogether convincing. The morbidity and mortality data for 'control' and study areas in 1996 did not reveal any clear benefits. The comparisons of morbidity rates made pre- and post-GAP revealed some benefits (those given above), but there was a lot of uncertainty and the same areas were not taken for the pre- and post-GAP samples. Hence, we need to look at these figures with some caution.

Assuming that they are correct, the loss of working days in the Ganga river area can be calculated as follows:

Gain in value of working days per user due to the GAP: Rs 73.57

Area within 0.5 km of river: 252 km^2

Average density: 500/km^2

Hence, total population affected: 12,62,500

Hence, total value of gain in working days: Rs 92.88 mn

This value has been taken for health benefits in the economic and financial analysis of the GAP in Chapter 12, subject to the caveats given there.

The other category of benefits estimated is for sewage farm workers. This works out to be Rs 4800 per worker per year. Assuming that there are 100 workers for each of the 35 GAP Sewage Treatment Plant schemes, this would amount to a total benefit of Rs 16.8 million. Again, this value has been added to the health benefits and used in the analysis of net benefits.

8

A Study of the Impact of the GAP Projects on Wastewater Toxicants and their Effects on Agriculture, Health, and the Environment

INTRODUCTION

As outlined in Chapter 1, part of the review of the GAP was to include an assessment of the impact of treated waste water (especially metals and pesticides) on environmental quality of the areas receiving the treated waste waters from the sewage treatment plants (STPs), as established under the Ganga Action Plan (GAP).

This chapter reports the results of that review. It was carried out by the Industrial Toxicology Research Centre (ITRC), under Dr Kunwar P. Singh as the principal investigator. The objectives of the study were:

1. Upgrading of the Ganga River Water Quality Database with particular reference to some selected metals and pesticides;
2. Assessment of the impact of treated/untreated waste water toxicants (metals and pesticides) discharged by the STPs under the GAP on public health, agriculture, and environmental quality in the disposal (receiving) areas.

The work carried out under the study consisted of the following:

1. Collection of Ganga river water samples from selected locations (out of 27 sites that had been identified under the GAP Phase I) and their analysis for selected metals and pesticides;
2. Identification and selection of sites in the areas receiving the treated/untreated waste waters under the GAP (exposed sites) and those not receiving such waters (unexposed sites). The latter

served as a control for the survey of agricultural, health, and environmental quality;

3. Collection of various environmental (water, soil, crops, vegetation, food grains, bovine milk, etc.) and biological (human blood and urine) samples from the treated wastewater receiving and non-receiving areas;

4. Collection of questionnaire-based information on health and agriculture status in the identified wastewater receiving and non-receiving areas through personal interviews;

5. Analysis of the collected samples of treated and untreated waste waters and of sludge from the STPs for general physical, chemical (including metals and pesticides), and bacteriological parameters. Analysis of environmental (water, soil, crops, vegetation, food grains, bovine milk etc.) and biological (human blood and urine) samples for metals and pesticides;

6. Analysis of soil samples collected from wastewater receiving and non-receiving areas for general characteristic parameters;

7. Statistical analysis of the collected information and analytical data generated with particular reference to metals and pesticides.

RIVER WATER QUALITY WITH REFERENCE TO SOME SELECTED METALS AND PESTICIDES

UPGRADING THE GANGA RIVER WATER QUALITY DATABASE

In order to upgrade the Ganga River Water Quality Database in relation to heavy metals and pesticides, so that comparison could be made with 1987 observations (pre-GAP), river water samples were collected from selected locations (as shown in Figure 8.1) on the river Ganga (viz. Kanpur up–stream or U/s, Kanpur down–stream D/s, Varanasi U/s, Varanasi D/s, Patna D/s, Rajmahal, Behrampur, Palta, Dakshineswar, and Uluberia) during the months of March to June 1996. These river water samples were processed and analysed for selected metals [cadmium (Cd), chromium (Cr), copper (Cu), iron (Fe), manganese (Mn), lead (Pb), nickel (Ni), zinc (Zn)], major organochlorine pesticides (BHC isomers, DDT metabolites and isomers, endosulfan), and organophosphorus pesticides (malathion, methyl parathion, ethion, dimethoate). The analytical data for pesticides and metals indicate a clear-cut decline in the metals and pesticide residue levels in the Ganga

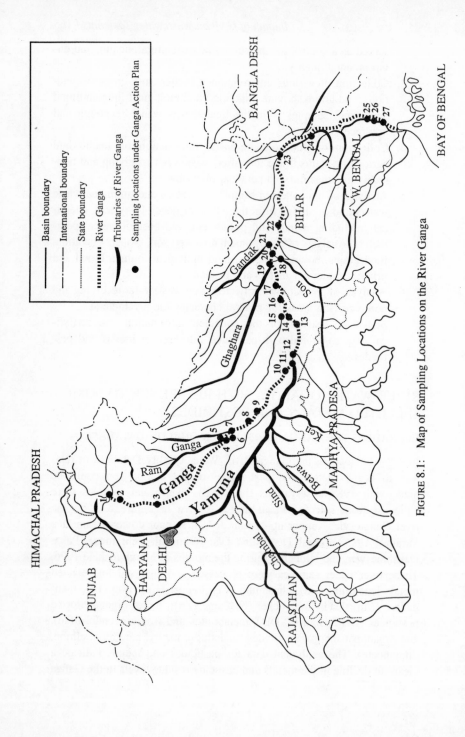

FIGURE 8.1:   Map of Sampling Locations on the River Ganga

river water at all these 10 locations when compared with those during 1987, the pre-GAP period.

Although it may not be directly attributable to the GAP, a significant decline in toxicant levels in the river water is an indirect result of the implementation of the GAP strategies in the river basin. Some of these strategic actions have included interception, diversion, and/or treatment of urban wastewater and industrial effluents in the catchment and also significant reduction in production of the organochlorine pesticides (DDT and BHC) during interceding years.

ASSESSMENT OF THE IMPACT OF TREATED/UNTREATED
WASTEWATER TOXICANTS (METALS AND PESTICIDES)
DISCHARGED BY THE SEWAGE TREATMENT PLANTS (STPs) ON
HEALTH, AGRICULTURE, AND ENVIRONMENTAL QUALITY
IN THE WASTEWATER DISPOSAL AREAS

*Identification and Selection of Sites in the Areas Receiving the
Wastewater (Exposed) under the GAP and Areas Not Receiving
(Unexposed) for Impact Assessment of the Wastewater
Toxicants (Metals and Pesticides) on Health, Agriculture,
and Environmental Quality*

Initially, it was planned that the treated wastewater impact analysis studies would be conducted in the areas (both receiving and not receiving wastewater) around Kanpur city. A Sewage Treatment Plant (UASB) of 5 MLD capacity has been established at Jajmau by the National River Conservation Directorate (NRCD). However, during the preliminary field survey it was noted that the treated waste water from the Jajmau UASB plant at Kanpur (5 MLD) is mixed up with the untreated wastewater and used for irrigation in the nearby adjoining outskirt area of the city. The total area receiving this wastewater for irrigation purpose is about 25 square kilometres. Along with this site near Kanpur city receiving the mixed treated/untreated wastewater, an adjoining area not receiving this wastewater was examined for comparative purposes.

*Assessment of the Toxicant (Metals and Pesticides) Loads
Discharged by the STPs*

The discharged treated wastewater from the Sewage Treatment Plants (STPs), is usually utilized for agricultural irrigation purposes in the nearby outskirt areas of the town and the generated sludge is disposed of through selling to horticultural or agricultural farmers. The treated wastewater and sludge from STPs, thus, on the one hand provides the benefits of irrigational support and manure value as nitrogen (N), phosphorus (P), potassium (K), organic carbon (C) etc., while on the other hand, both these components may be contaminated with high levels of the toxicants (such as heavy metals and pesticides etc.). In the long run this may contaminate the surface and groundwaters, soils, crops and vegetation, and food, especially fruit and vegetables, causing considerable adverse impacts on the health of the consumers/local population as a result of environmental exposure, and at the same time, affect the agricultural crops and the environmental quality in the disposal areas. Therefore, to analyse and evaluate the impacts of treated wastewater toxicants, it is important to, firstly, assess and estimate the quantum of these toxicants, loadings to the study area, which obviously requires the characterization of both, the treated wastewater and sludge generated by the STPs for the parameters of interest, such as levels of these heavy metals and pesticides.

TABLE 8.1

Villages Selected for Impact Analysis for the Wastewater Toxicants

Site	Village	
	Receiving wastewater (exposed)	*Not receiving wastewater (unexposed)*
1. Kanpur	Shekhpur	Paligaon
	Motipur	Kurgaon
	Kishanpur	Chakeri
	Madarpur	T Pagambarpur
	Hannia	Gauria
	Trilokpur	C Chhatimara
	Gadarianpurva	B Chhatimara
	Karvigaon	
2. Varanasi	Dinapur	Tatepur
	Kotawa	Kapildhara
	Danipura	Khalispur
	Kamauli	
	Nawapura	

CHARACTERIZATION OF UNTREATED/TREATED
WASTEWATER/SLUDGE (STPS)

Wastewater (Untreated/Treated)

Raw (untreated) and treated wastewaters samples were collected
from the 'inlet' and 'outlet' of the Sewage Treatment Plants (STPs) in
Jajmau, Kanpur (5 MLD) and Dinapur, Varanasi (80 MLD) between
April and October 1996 during the peak (morning and evening) and
non-peak (noon) hours to assess treated and untreated wastewater
quality and work out the quantitative loading of the toxicants in the
receiving area.

The analytical data (mean) pertaining to the inlet (untreated) and outlet (treated) wastewaters of the two Sewage Treatment Plants at Kanpur
and Varanasi with particular reference to the metal and pesticide residue
levels are shown in Tables 8.2 and 8.3.

TABLE 8.2
Mean Metal Levels in Wastewater

(mg/l)

STP	Cd	Cr	Cu	Fe	Mn	Ni	Pb	Zn
Jajmau, Kanpur (UASB, 5 MLD):								
Inlet	0.05	6.45	0.88	8.80	0.55	0.22	0.19	1.82
Outlet	0.01	0.37	0.03	0.42	0.17	0.05	0.05	0.14
Dinapur, Varanasi (STD, 80 MLD):.								
Inlet	0.16	8.12	0.16	3.32	0.47	0.14	0.15	1.58
Outlet	0.02	0.02	0.03	0.30	0.14	0.05	0.05	0.12
WQ Crit Irrigation	0.01	0.1	0.2	5.0	0.2	0.2	5.0	2.0

TABLE 8.3
Mean Pesticide Residue Levels in Wastewater

(mg/l)

STP	a-BHC	B-BHC	r-BHC	T-BHC	pp-DDT	pp-DDD	pp-DDE	T-DDT
Jajmau, Kanpur (UASB, 5 MLD):								
Inlet	0.11	0.095	0.027	0.232	ND	0.02	2.25	2.51
Outlet	0.027	0.016	0.009	0.051	ND	0.002	0.55	0.56
Dinapur, Varanasi (STD, 80 MLD):								
Inlet	0.098	0.038	0.030	0.166	ND	ND	0.121	0.132
Outlet	0.018	0.009	0.008	0.035	ND	ND	0.021	0.024

From the mean levels of various toxicants (metals and pesticides) in the treated (outlet) wastewaters of the two STPs, their total input/discharge and loading of these pollutants can be estimated, as received in the respective disposal/application areas under study (see Table 8.4).

TABLE 8.4
Mean Metal Discharge through Treated Wastewater

STP	Cd	Cr	Cu	Fe	Mn	Ni	Pb	Zn
Kanpur:								
(g/d)	0.06	1.84	0.16	2.08	0.83	0.25	0.27	0.67
(kg/yr)	0.02	0.67	0.06	0.76	0.30	0.09	0.097	0.25
Varanasi:								
(g/d)	1.90	1.62	2.57	28.5	13.5	4.47	4.56	11.60
(kg/yr)	0.694	0.59	0.94	10.4	4.92	1.63	1.66	4.23

SUITABILITY OF TREATED WASTEWATER FOR IRRIGATION: As far as the suitability of the treated wastewaters (with particular reference to the heavy metals) for irrigation of agricultural land is concerned, it can be noted from Table 8.2 that apart from the cadmium levels in treated water from the two STPs and the chromium levels in the Jajmau (Kanpur) STP, the levels of all the other metals are within the permissible limits as prescribed for agricultural land irrigation purposes. Further, the higher levels of nitrogen (N), phosphorus (P), and potassium (K) in the treated wastewater result in a positive impact on crop productivity in the receiving areas.

TABLE 8.5
Pesticide Discharge through Treated Wastewater

STP	r-BHC (lindane)	Total DDT
Kanpur:		
(g/d)	0.045	2.85
(5 MLD) (kg/yr)	0.016	1.04
Varanasi		
(g/d)	0.760	2.28
(80 MLD) (kg/yr)	0.277	0.83

STPs Sludge

STPs sludge samples were also collected regularly from both the treatment plants and analysed for heavy metals, pesticides residue, and the N, P, K, and organic C contents. The results (mean values) are presented in Tables 8.6, 8.7, and 8.8.

TABLE 8.6
Mean Metal Levels in STPs Sludge

(g/kg)

STP	Cd	Cr	Cu	Fe	Mn	Ni	Pb	Zn
Kanpur	0.041	8.11	0.393	6.40	0.22	0.214	0.091	1.18
Varanasi	0.054	1.30	0.543	7.21	0.31	0.293	0.129	1.51
Tolerable amounts*	0.02	1.20	1.20	–	–	0.20	–	3.00

*FRG: Federal Biological Research Centre for Agriculture and Forestry

TABLE 8.7
Mean Pesticide Levels in STP Sludge

(mg/kg)

STP	a-BHC	B-BHC	r-BHC	T-BHC	pp-DDT	pp-DDD	pp-DDE	T-DDT
Kanpur	0.07	0.44	0.36	0.69	0.09	0.05	0.12	0.24
Varanasi	0.07	0.08	0.05	0.19	0.09	0.03	0.02	0.11

TABLE 8.8
Mean N, P, K, and OC Contents in STPs Sludge

(g/kg)

STP	TKN	Tot Phosphate (PO_4-P)	K	Org C
Kanpur	13.10	7.25	4.54	661.12
Varanasi	14.89	6.32	4.00	552.28

From Table 8.6, it is clear that sludge generated by both the STPs have cadmium, chromium, and nickel levels at about their tolerable amounts as prescribed for agriculture land application.

However, in terms of the fertilizer value (N, P, K, etc.) from Table 8.8, it can be estimated that about 14 kilograms nitrogen (N), 6.75 kilograms phosphorus (P), and 4.25 kilograms potassium (K) would be available per ton of the generated sludge. Their economic value at existing market rates are given below in Table 8.9.

TABLE 8.9

Mean N, P, and K Contents of STPs Sludge and its Cost Economics

Ingredient	Rate per kg (Rs)	Per ton of sludge	
		Ingredient (kg)	Cost (Rs)
Nitrogen (N)	7.6	14.0	107
Phosphorus (P)	37.0	6.75	243
Potassium (K)	32.0	4.25	136

From Table 8.9, it may be noted that the cost of fertilizer ingredients (N, P, K) per ton of the STPs sludge generated is about Rs 486.

IMPACT OF WASTEWATER TOXICANTS ON ENVIRONMENTAL QUALITY OF DISPOSAL AREAS

METAL AND PESTICIDE RESIDUE LEVELS IN THE ENVIRONMENT

To assess the impact of the wastewater/sludge disposal (metals and pesticides) on the environmental quality of the receiving/application areas around the respective STPs, environmental samples (surface water, groundwater, soil, vegetables, crops, food grains, milk etc.) were collected from different villages, both in the receiving (exposed) and non-receiving (unexposed) areas identified and finally selected near UASB, Jajmau, Kanpur and STP, Dinapur, Varanasi. The statistics of samples collected are as below.

These environmental samples (water, soil, crop/vegetation and food grains etc.) collected from the treated/untreated wastewater receiving and non-receiving areas around the UASB Kanpur and STP Varanasi were processed and analysed for some selected heavy metals (Cd, Cr, Cu, Fe, Mn, Pb, Ni, and Zn) and major pesticide (BHC isomers, DDT isomers and metabolites, endosulfan, malathion, methyl parathion, ethion and dimethoate etc.) residue levels. Details of the number of samples collected in the different areas are given in Table 8.10.

From these data, it may be noted that the level of heavy metals as well as pesticides (BHC and DDT) in each of the environmental media (water, soil, vegetation/foodgrains etc.) is much higher in the area receiving the waste water for irrigation purposes as compared to the adjoining area where there is no irrigation with this water. Further, it may be noted that in some of the cases the levels of groundwater are not much different. This may be due to common groundwater aquifers in both the areas. Surface water data collected from only a few available ponds show higher levels of metals as well as pesticides. This water is usually used as drinking water for cattle. Metal and pesticide levels in soils from waste water disposal areas are also much higher than those from the respective unexposed areas. This might have resulted from the continued application of the wastewater. Metal and pesticide residue levels in the vegetation/crops, vegetables/foodgrains etc., as consumed by cattle and human beings, are also much higher in the waste water disposal areas than those not receiving the wastewater. However, the higher pesticide residue levels in different environmental media may not solely be due to the waste water discharge as there are several other uses and applications of pesticides such as crop protection and health programmes. The soil samples were also analysed for their characteristic physico-chemical parameters (pH, densities, pore space, electrical conductance, nitrogen (N), phosphorus (P), potassium (K), organic carbon (C), calcium (Ca), magnesium (Mg), and sodium (Na) contents etc.).

TABLE 8.10

Statistics of the Samples Collected from Sites Selected near the Kanpur and Varanasi STPs

Environmental media	Kanpur (no. of samples)		Varanasi (no. of samples)	
	Receiving (exposed)	Non-receiving (unexposed)	Receiving (exposed)	Non-receiving (exposed)
Water	18	13	14	13
Soil	8	7	7	6
Vegetables/Crops	13	7	7	7
Foodgrains	11	7	6	5
Total	*50*	*34*	*34*	*31*

The heavy metal and pesticide residue levels in the soils, both near Kanpur and Varanasi, irrigated with waste water from the STPs were

TABLE 8.11
Characteristics of Soil near Kanpur

	pH (1:5)	EC (1:10) (μmho/cm)	Bulk density (g/cc)	Particle density (g/cc)	Pore space (%)	Org C (%)	Tot-N (ug/g)	P (ug/g)	Na (ug/g)	K (ug/g)	Ca (mg/g)	Mg (mg/g)
Kanpur Exposed Area:												
Range	7.5–8.9	1050–1250	0.96–1.63	1.54–2.50	19.44–50.74	1.12–2.36	392.0–1120.0	4.95–7.95	58.80–305.6	38.4–130.0	0.89–4.60	0.34–1.77
Mean	8.1	1160	1.22	1.88	35.56	1.50	738.5	5.96	148.3	93.7	3.36	0.81
Kanpur Unexposed Area:												
Range	7.9–9.0	1050–1250	0.93–1.35	1.82–2.22	31.32–49.64	0.62–1.49	56.0–1120.0	1.74–8.47	26.6–137.2	28.8–49.6	1.39–3.39	0.31–1.36
Mean	8.71	1130	1.19	1.97	39.63	1.05	556.0	4.16	50.6	34.86	2.47	0.80

TABLE 8.12
Characteristics of Soil near Varanasi

	pH (1:5)	EC (1:10) (umho/cm)	Bulk Density (g/cc)	Particle Density (g/cc)	Pore Space (%)	Org C (%)	Tot-N (ug/g)	P (ug/g)	Na (ug/g)	K (ug/g)
Varanasi Exposed Area:										
Range	7.0–8.5	900–1150	0.88–1.48	0.92–2.0	3.94–41.85	0.14–0.51	765–1428	3.45–10.94	58.8–104	52.8–114
Mean	8.09	1030	1.21	1.72	26.65	0.35	1108	6.36	78.8	83.74
Kanpur Unexposed Area:										
Range	8.2–8.55	825–1050	1.01–1.25	1.79–2.0	29.86–49.7	0.29–0.47	840–1400	8.21–13.56	70.4–99.4	76–114
Mean	8.43	995	1.16	1.9	38.51	0.37	1071	10.76	87.66	97.6

found to be higher compared to those in adjoining soils not receiving the wastewater. Further, the soil characteristics results (see Table 8.11 and 8.12) indicate a considerable alteration in the soil quality of the areas irrigated with the wastewater over the ones not receiving the wastewater.

IMPACT OF WASTEWATER TOXICANTS

IMPACT OF WASTEWATER TOXICANTS ON HEALTH

Both the untreated and treated wastewaters may contain a variety of inorganic and organic chemicals. Among these, heavy metals and pesticides are more commonly found to be present. Exposure to these hazardous chemicals may be exhibited by several signs and symptoms (Otto *et al.*, 1994; Raymond *et al.*, 1995; Williams, 1996) but they are only recognized when they achieve chronic and clinical levels. Since the pesticides and heavy metals are both proven neurotoxic substances, there are several methods developed and reported (Anger and Kent, 1989; Hanninen and Lindstrom, 1979) to assess exposure at very low concentrations of these substances (sub-clinical level). Further, under these conditions, the signs and symptoms which may appear and be observed are of a reversible nature; neurobehavioural assessment of the expected exposed population along with biomonitoring components establishing the toxicant levels in biological components (human blood and urine) is of much significance in the drawing of meaningful inferences.

Since the exposure of the population to heavy metals and pesticides may cause neuro-behavioural disorders such as fatigue, insomnia, decreased concentration, depression, irritability, gastric symptoms, sensory symptoms, and motor symptoms etc., a chemicals (metals and pesticides) based symptoms standard questionnaire (Hanninen and Lindstrom, 1979) was modified, which comprised a total of 35 items covering eight different functions for surveying and recording information from selected populations with regard to any impact of these toxicants (heavy metals and pesticides) (Proforma 1). This was used during the survey through personal interviews, both in the exposed and unexposed populations near Kanpur and Varanasi towns. Neurobehavioural analysis was carried out on the basis of mean scores for each function of every individual. Table 8.13 describes the neurobehavioural functions that were covered in the questionnaire.

TABLE 8.13

Neurobehavioural Functions Covered in the Questionnaire

Function	Number of items	Possible score range
Fatigue	4	0–4
Insomnia	3	0–3
Decreased Concentration	4	0–4
Depression	5	0–5
Irritability	6	0–6
Gastric Symptoms	4	0–4
Sensory Symptoms	4	0–4
Motor Symptoms	5	0–5

Representative population groups in all the four areas (near Kanpur and Varanasi) were surveyed for general health status and neurobehavioural functions (Proforma 1) through questionnaire-based personal interviews. Table 8.14 gives the numbers interviewed in each area as part of this survey.

TABLE 8.14

Exposed and Unexposed Population Groups Surveyed for Health and Neurobehavioural Symptoms

Area	Number of persons interviewed	
	Exposed	Unexposed
Kanpur	53	52
Varanasi	51	50

The neurobehavioural information (Proforma 1) was analysed statistically applying the Student's 't' test to see the significant difference in the mean of different parameters between the unexposed and exposed population groups near Kanpur and Varanasi Sewage Treatment Plants (STPs) separately and the results (significant/insignificant over their respective unexposed population groups) indicating level of significance (p) are presented in Table 8.15.

From Table 8.15, it may be noted that overall, and for the individual neurobehavioural function, the differences in the population selected from the area receiving waste water (mixed) from the UASB, Jajmau, Kanpur (exposed) were significant over the respective unexposed population groups selected from the adjoining area, but receiving no wastewater. Further, the differences among Varanasi population groups were non-significant for all the functions. However, this does not mean that there is no significant exposure to heavy metals and pesticides or risk to the population near the Varanasi area receiving treated wastewater from the Dinapur STP (80 MLD). One of the possible reasons seems to be the duration of disposal, as the area near Kanpur had been receiving wastewater for several decades, while the one near Varanasi started receiving it only during the last few years. This is further supported by the observed higher levels of metals and pesticides in different environmental compartments in the disposal area near Kanpur as compared with those near Varanasi.

TABLE 8.15

Analysis of the Neurobehavioural Functions in Population Groups (Exposed and Unexposed) near the Kanpur and Varanasi STPs

Function	Kanpur	Varanasi
Fatigue	+	–
Insomnia	+	–
Decreased concentration	+++	–
Depression	++	–
Irritability	++	–
Gastric symptoms	+++	–
Sensory symptoms	++	–
Motor symptoms	+	–

(+) Significant at $p < 0.05$; (++) at $p < 0.01$; (+++) at $p < 0.0001$. (–) Non-significant

Since both heavy metals and pesticides are very persistent, these may accumulate through their long run disposal. Further, the extent of exposure of the selected populations to the elevated levels of toxicants in different environmental media was assessed through biological monitoring.

BIOMONITORING OF THE METALS AND PESTICIDES LEVELS (HUMAN BLOOD AND URINE) IN EXPOSED/UNEXPOSED POPULATIONS

In order to assess further the extent of the metals and pesticides exposure of the population in the area receiving the wastewater, the levels of various metals and pesticides were measured in human blood and urine samples. Intravenous blood samples of about 10 millilitres and urine (24 hours) samples were collected in all four areas from the representative population groups. These samples were immediately transported to ITRC under low temperature conditions (in ice boxes) and were processed for analysis of heavy metals and pesticides separately as per standard methods. Table 8.16 gives the statistics of the samples collected.

TABLE 8.16
Statistics of Samples for Biological Monitoring

Area	Number of blood samples		Number of urine samples	
	Exposed	*Unexposed*	*Exposed*	*Unexposed*
Kanpur	12	12	38	30
Varanasi	14	16	32	22

The analytical results indicate that the metal as well as pesticide levels (mean) in blood and urine samples collected from the population representing wastewater irrigated areas were higher than those collected from the areas not receiving the wastewater in the case of both Kanpur and Varanasi. Further, it may be noted that the residue levels of metals and pesticides in human blood as well as in urine samples of the exposed and unexposed population groups in Kanpur were higher as compared with those of Varanasi exposed and unexposed population groups, respectively. This may again be due to the prolonged disposal of the wastewater in the area near Kanpur over decades, resulting in long term exposure of the population.

ENVIRONMENTAL EXPOSURE RISK ASSESSMENT OF METAL AND PESTICIDE

The environmental exposure risk to the populations from these elevated levels of metals and pesticides in different environmental media (water, food, vegetables, crops/vegetation, milk, etc.) in areas receiving wastewaters over unexposed ones has been evaluated by first computing the mean estimated Total Daily Intake (TDI) of each of these toxicants (individual metal and pesticide) as

$$\text{TDI (mg/day)} = \Sigma \, C_i \cdot D_i \qquad (8.1)$$

where C_i is the mean concentration of individual toxicant in the ith media and D_i is the mean daily intake of the same media by a person. The major intake routes considered are: drinking water (2.5 l/d), food-grains (600 g/d), vegetables (300 g/d), and milk (200 g/d).

The computed TDI (mg/d) values for each toxicant are then compared with their respective Acceptable Daily Intake (ADI) values (mg/d), worked out from their individual ADIs (mg/d/kg b.wt) as available in the literature for a person of 60 kilograms body weight.

The risk quotient (RQ) for each toxicant has been computed as

$$RQ = TDI/ADI. \qquad (8.2)$$

As a general principle, the population exposed to some particular toxicant (chemical) will be at risk with respect to that toxicant if the value of the respective risk quotient (RQ) is above 1.0. However, a comparison of the respective RQs of two population groups with respect to some common toxicant to which these are exposed may give an assessment of their relative risk level for that particular toxicant.

However, none of the populations is at a significant exposure risk level, as in none of the cases does the computed RQ value (TDI/ADI) exceed 1.0. However, it is very clear that the exposure risk level of the exposed population groups (Kanpur and Varanasi) with respect to each of these metals and pesticides is much higher (2 to 4 times) as compared with the respective unexposed population groups.

IMPACT OF WASTEWATER TOXICANTS (METALS AND PESTICIDES) ON AGRICULTURE

The treated wastewater and sludge containing elevated levels of persistent toxicants such as metals and, to some extent, pesticides, in the long run disposal for irrigation or amendment of agricultural land may lead to a building up of higher concentrations in soils as a result of accumulation. The higher metal levels in soil may cause a negative impact on crops, inhibiting the growth in one way or another. One of the most important factors is the pH of the soils. Alkaline pH of the soil would usually restrict the mobilization of the metal in the soil matrix and consequently, the metal uptake by the crop plant would be controlled, obviously reducing the risk of metal toxicity.

The critical concentrations of various metals in agricultural soils tolerable to crops are reported as: Cd (3 mg/g); Cr (100 mg/g); Cu (100 mg/g); Ni (50 mg/g); Pb (100 mg/g); Zn (300 mg/g). From data collected by the Institute carrying out this research, it may be noted that the mean levels of Cd and Cr were found to be above their critical levels in agricultural soils of the area near Kanpur STP irrigated with wastewater. The mean levels of Cd, Ni, and Pb in soils of treated wastewater irrigated area near Varanasi STP were above their respective tolerable limits for agricultural crops. However, the disposed wastewater in both the areas had a mean pH value of about 8 and the mean pH value for the agricultural soils irrigated with the wastewater was found to be more than 8. Therefore, even though the level of a few metals in soils were above their critical limits, their mobilization and plant uptake appears to be restricted by the alkaline pH.

Further, it may be mentioned that the critical levels of the heavy metals in soils displaying negative impacts on agricultural crops are high relative to the actual level found in the study areas irrigated with the treated wastewater. Therefore, as yet, there seems no adverse impact of metals and pesticides on agricultural crops in these areas. A questionnaire-based individual farmers' survey was conducted in all the four exposed and unexposed areas near Kanpur and Varanasi STPs, collecting information on agriculture crops production trends during the last few years. In the exposed areas near Kanpur, the majority (90 per cent) reported that the crop yield has declined over the past few years due to some root disease infestation causing plant death or weakness, leading to small grain size. But in the area near Varanasi irrigated with treated wastewater, the response of the farmers was totally the reverse. An enhancement in crop yield over the last few years was reported by the majority of the farmers (65 per cent). The reason for declining productivity near Kanpur may not be due to the toxic wastewater carrying high solids/bacterial biomass, resulting in deposition in soils, as observed during the survey and making the soil–root interface more susceptible to plant root diseases. The enhanced yield in the area near Varanasi irrigated with treated wastewater may be accounted for by more irrigation water availability with a high nutrient/fertilizer (N, P, K, organic C, etc.) value to the crops as discharged by the Dinapur STP.

Further, in a study recently completed at ITRC on utilisation of fly ash in agriculture, mainly establishing the impact of the elevated heavy metal level in soils with fly ash added on agricultural crops, it has been concluded that despite the very high levels of heavy metals in fly ash mixed with the soils, there was virtually no impact on agricultural crops in terms of their growth and yield. There was also no adverse impact on crop quality assessed in terms of their lipid, protein, and carbohydrate contents.

CONCLUSIONS

The major conclusions drawn from the above study are as follows:

UPGRADING OF THE GANGA RIVER WATER QUALITY DATABASE

The levels of different heavy metals (Cd, Cr, Cu, Fe, Mn, Pb, Ni, Zn) and pesticides (BHC isomers; DDT isomers/metabolites and endosulfan etc.) in the Ganga river water monitored during 1996 (post-GAP Phase-I) at 10 selected but common locations identified earlier under the Ganga Action Plan (Kanpur u/s, Kanpur d/s, Varanasi u/s, Varanasi d/s, Patna d/s, Rajmahal, Behrampore, Palta, Dakshineswar, and Uluberia) were much lower as compared with those observed during the year 1987 (pre-GAP). None of the organophosphorus pesticides (malathion, methyl parathion, ethion, dimethoate) were detected in the river water during 1996. Although the decline in the level of metals and pesticides is not directly related to the GAP strategies adopted under the programme, it is, however, an indirect impact of several GAP strategies implemented in the river basin.

IMPACTS OF WASTEWATER TOXICANTS (METALS AND PESTICIDES)

Since the conventional type STPs are basically designed to reduce the organic load, these are not very effective in reducing the levels of metals and pesticides etc., except that a larger fraction of these toxicants present in the wastewater is retained with the sludge generated by STPs, the remainder going out with the treated wastewater/effluents.

ON ENVIRONMENTAL QUALITY

The impact of these treated wastewater toxicants (metals and pesticides) on the environmental quality of the disposal areas, as assessed in terms of their elevated levels in different media samples, viz. water, soil, crops, vegetation, foodgrains, and biological samples collected from exposed areas, over the respective unexposed areas (not receiving wastewater) near Kanpur and Varanasi STPs indicates that as a result of long term disposal of these toxicants, high levels build up and will obviously be hazardous to the population. Since both metals and organochlorine pesticides are of a persistent type, staying long in the environment, higher levels of these chemicals are built up in the long term.

The analytical data generated on metals and pesticide levels in various environmental media in both exposed as well as unexposed areas shows their elevated levels in all the environmental compartments in exposed areas as compared to those in unexposed ones. Therefore, these toxicants have a definite adverse impact on the environmental quality of the disposal areas.

ON HEALTH

The impact of waste water toxicants (metals and pesticides) on human health in the areas receiving wastewater was assessed through a standard questionnaire-based survey of the exposed and unexposed population groups near Kanpur and Varanasi STPs. The questionnaire contained a total of 35 items which covered eight neurobehavioural functions established to be affected by chemical (heavy metals and pesticides) exposure. Neurobehavioural analysis was carried out on the basis of mean scores for each function of every individual. This statistical analysis for overall and function-wise differences between the unexposed and exposed population groups indicated a significant difference between the two groups near Kanpur. Therefore, there has been a considerable impact of these toxicants (metals and pesticides) on human health in the exposed areas.

Further, an Environmental Exposure Risk Analysis for these four population groups (two exposed and two unexposed) was carried out for each of these toxicants individually. The approach was based on evaluation of the Risk Quotient (RQ) for each individual toxicant by first computing the Total Daily Intake (TDI) of each one through the major routes (drinking water, food grains, vegetables, milk, etc.) and then comparing with the respective Acceptable Daily Intake (ADI). The final values of RQs indicated that although in none of these cases did the RQ value exceed 1.0 (positive risk), the RQ values for all the metals and pesticides for the two exposed areas were 2 to 4 times higher than their respective unexposed population groups. This also supports the premise that there is considerable risk through metal and pesticide exposure to human health.

The impacts were further confirmed through biomonitoring of the metal and pesticide levels in human blood and urine of the different population groups under study. The levels of both metals and pesticides in the human blood and urine samples of the two exposed population groups (Kanpur and Varanasi) were considerably higher than those of the respective unexposed population groups. Thus, all the three different approaches indicated a considerable risk and impact of heavy metals and pesticides on human health in the exposed areas receiving the waste water from STPs.

ON AGRICULTURE

It may be noted that the mean level of Cd and Cr in soils near Kanpur and Cd, Ni, and Pb in soils near Varanasi are above their respective tolerable limits for agricultural crops. However, since the pH of the waste waters as well as the receiving soils is more than 8, the metal mobilization and plant uptake would be restricted by the alkaline pH.

Since the critical levels of the heavy metals in the soil for agricultural crops are much higher than those observed in our study areas irrigated with waste water, there seems no adverse impact of metals and pesticides on agricultural crops in these areas. But questionnaire-based information on agriculture crops' yield during the last few years, collected from these areas, revealed that the crops' yield has declined (90 per cent cases) over the past few years, while an enhanced yield was reported (65 per cent) in the area near Varanasi irrigated with the treated waste water. The decreased productivity in the previous case was due to high solids/bacterial biomass making the soil–root interface more susceptible to plant root diseases. The enhanced yield in the latter case may be due to more irrigation water available with high nutrient/fertilizer (N, P, K, organic C, etc.) levels.

IMPACT OF STPS SLUDGE

The STP's sludge has both positive and negative impacts as it is enriched with high levels of toxic heavy metals and pesticides, but also with several useful ingredients such as N, P, K, etc., which provide fertilizer values. The STPs' sludge studied has cadmium, chromium, and nickel levels above their tolerable levels as prescribed for agriculture land application. However, the soil pH (mean) is in the alkaline range (>8.0), counteracting metal mobilization and restricting uptake by the crops to a considerable extent, thus reducing the expected toxicity.

In terms of the fertilizer value (N, P, K, etc.), it has been estimated that approximately 14 kilograms of nitrogen (N), 6.75 kilograms of phosphorus (P), and 4.25 kilograms of potassium (K) would be available per ton of the generated sludge and its value at existing market rates would be be about Rs 486.

9

Fisheries on the River Ganga

INTRODUCTION

Fisheries are an important resource for the people who live close to the banks of the river Ganga. Estimates of total catch are difficult to make, but in 1995 reported catches were nearly 400 tonnes. This is an underestimate of the total catch, as there are innumerable points where catches are landed and disposed off, and where no records are kept.

This study is interested in ascertaining whether the GAP has had any effect on fish catch, when allowance has been made for other factors, particularly the level of effort and the quality of equipment. To this end, the Central Inland Capture Fisheries Research Institute was asked to collect data on:

(a) Fishery locations: where fish is landed;
(b) Species caught and catch volume;
(c) Catch effort;
(d) Commercial Value;
(e) Significant non-GAP changes that could affect fish catch;
(f) Significant GAP changes that could affect fish catch.

This chapter is based on a report by ICAR (ICAR, 1997), and reports on the main findings on total landings and trends in those landings, as far as can be established. From the point of view of riverine fisheries, there are two stretches where most of the catch is made: one is the section from Kanpur to Farakka and the other is the lower stretch from Farakka to Diamond Harbour. The data are reported for both sections and for major landing points in each section. It then goes on to examine trends in total effort and catch per unit of effort and discusses the impact of water quality changes on fish populations. It then draws some overall conclusions on changes in fisheries in the river over the period 1985–95 and what can be said of the importance of the GAP in bringing about these changes.

FISH CATCH AND TOTAL LANDINGS: TRENDS SINCE 1985

Data are available for catch by species for the following landing stations:

- Sadiapur, Daraganj, Patna, Buxar, Bhagalpur, and Lalgola (above Farakka)
- Medgachi, Balaghar, Hooghly Ghat, Nawabganj, Konnagar, Baranagar, Uluberia, Godakhali, and Nurpur (below Farakka).

Data are summarized for total catch in Tables 9.1 and 9.2 and are shown graphically in Figures 9.1 and 9.2. As can be seen, there are considerable data gaps for the stations above Farakka. Moreover, some of the stations, for which there are no recent data, have been quite important landing points (for example Patna and Bhagalpur).

TABLE 9.1
Total Fish Catch above Farakka

(tonnes)

Year	Landing Centre					
	Sadiapur	Daraganj	Patna	Buxar	Bhagalpur	Lalgola (*)
1985	132.93			26.75	62.54	
1986	135.82		69.49	41.15	66.77	
1987	113.58	32.76	62.12		68.89	
1988	100.93	28.03	55.73		50.23	
1989	84.92	25.44	43.08			48.03
1990	71.32	22.24	44.53			71.42
1991	90.99	23.66	32.58			73.66
1992	63.80	26.20	35.14			75.81
1993	46.27	25.85	33.61			93.93
1994	68.51	29.60				81.06
1995	55.27	21.93				133.52

Note: (*) Figure for Lalgola in 1989 is partially an estimate.

The data for stations above Farakka reveal that fish landings at Sadiapur (Allahabad) and Patna have consistently declined, whereas landings in Lalgola have gone up sharply. Other centres show no dis-

TABLE 9.2
Total Fish Catch below Farakka

(tonnes)

Year	Landing Centre										Total
	Medgachi	Balaagarh	Hooghly Ghat	Nawabganj	Konnagar	Baranagar	Uluberia	Godakhali	Nurpur		
1985	8.04	10.43	11.55	8.14	5.8	19.69	16.69	43.09	31.13	154.56
1986	6.71	13.52	10.73	6.64	3.28	16.63	15.72	43.99	58.4	175.62
1987	8.81	24.28	15.04	11.34	4.77	21.16	18.86	60.38	79.37	244.01
1988	8.87	17.26	11.3	12.64	4.88	21.98	15.24	25.32	28.97	146.46
1989	6.8	9.34	9.78	8.89	3.67	16.07	18.23	54.39	14.27	141.44
1990	8.53	9.93	13.1	20.44	3.33	20.57	17.97	47.98	16.86	158.71
1991	7.21	9.98	10.81	19.49	3.72	20.23	15.63	42.86	27.62	157.55
1992	9.14	10.94	13.34	13.8	2.97	19.82	20.97	35.65	28.6	155.23
1993	12.9	10.12	10.45	9.22	3.99	14.61	18.23	33.63	37.52	150.67
1994	8.68	11.44	14.78	12.13	3.97	16.79	24.91	61.51	38.02	192.23
1995	8.71	11.9	13.2	10.88	4.01	14.26	14.99	61.46	49.18	188.59

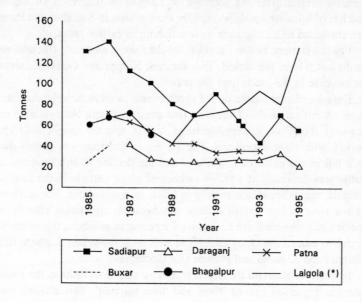

FIGURE 9.1: Landings at Different Centres Above Farakka

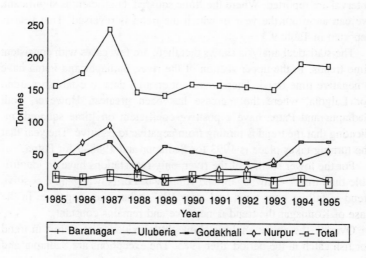

FIGURE 9.2: Landings at Different Centres Below Farakka

cernible trends. The big increase at Lalgola is the result of higher catches of hilsa, an anodromous fish; the declines in Sadiapur and Patna are attributed to factors such as overfishing in earlier years.

For the stations below Farakka, the data show moderate fluctuations in the catch over the period. Two stations, Nurpur and Godkhali, show an increase in the catch over the period.

Data are also available on the species-wise catch at these landing stations. A critical study of the data (not presented here but available on request) shows no discernible trend in fish species caught over the period, with some exceptions. The decline in Sadiapur was mainly due to a fall in the catch of *L. calbasu* and the decline in fish landings at Patna was the result of a fall in catches of major catfish. In the case of Lalgola, the increase was mainly in hilsa, as noted above. For sections of the river below Farakka, there has been no significant change in species mix. Where there has been an increase in numbers, this is mainly due to more hilsa being caught, as well as *S. pama* and *S. phasa* (for Nurpur) and *S. pama* and prawns (for Godakhali).

In order to ascertain the trends in a more statistical fashion, the catch rate was regressed against 'time' and 'time squared'. This allowed one to see if there was a trend and if the trend had changed over the period. The results of that exercise are reported in Table 9.3. Only cases where the coefficients are statistically significant at the 95 per cent confidence interval are reported. Where the 'time squared' coefficient is significant, we can ascertain the year in which the trend is reversed. That year is reported in Table 9.3

The statistical analysis shows that there are few cases with consistent time trends. In the upper section of the river, Sadiapur and Patna have a negative time trend. There were not enough data to establish a trend for Lalgola, where the increase has been greatest. However, both Sadiapur and Patna have a positive coefficient on 'time squared' indicating that the trend is turning from negative to positive. The year that the turning takes place is 1993 for Sadiapur and 1988–9 for Patna.

For the lower section of the river, only two stations have an identifiable time trend—Nawabganj and Konnagar. For Nawabganj, a positive trend is falling over time, and after 1987 it becomes negative. In the case of Konnagar the trend is negative and remains constant.

Overall, the above analysis shows little evidence of a shift in trend for fish catch in the period after 1985. The exceptions are Sadiapur and Patna, where a negative trend appears to have been reversed, and Nawabganj where a positive trend appears to have been reversed, and

Konnagar, where a negative trend is identified. Links of these trends to the GAP remain uncertain and problematic.

TABLE 9.3

Trends in Catch Landed at Various Stations: 1985–95

Landing station	Coefficient on time	Coefficient on time squared	Adjusted 'R' squared	Turning year
Stations above Farakka:				
Sadiapur	−16.9	+0.70	87.0%	1993
Daraganj	n.s.	n.s.	–	–
Patna	−13.0	+0.69	96.1%	end 1988
Buxar	lack of data	lack of data	lack of data	lack of data
Bhagalpur	lack of data	lack of data	lack of data	lack of data
Lalgola	lack of data	lack of data	lack of data	lack of data
Stations below Farakka:				
Medgachi	n.s.	n.s.	–	–
Balagarh	n.s.	n.s.	–	–
Hooghly Ghat	n.s.	n.s.	–	–
Nawabganj	+3.76	−0.28	43.0%	1987
Konnagar	−0.58	n.s.	43.0%	–
Baranagar	n.s.	n.s.	–	–
Uluberia	n.s.	n.s.	–	–
Godakhali	n.s.	n.s.	–	–
Nurpur	n.s.	n.s.	–	–
Diamond Harbour	n.s.	n.s.	–	–

TRENDS IN CATCH PER UNIT OF EFFORT AND TOTAL EFFORT

The Ganga has a multi-species fishery which is exploited by a variety of gears and nets of various mesh and size. The prominent gears used in the middle stretch are gill-nets, followed by long lines and dragnets. The landings are mostly recorded from assembly centres and hence, the gear-wise catch in this stretch is not available. On the lower stretch of the river the type of gear used includes: trawl, small seine, purse seine, drift-gill-net, lift, cast, and hook and line. These are of different sizes depending on the area of operation. Landings data on the lower stretch are collected on a gear-wise basis. This allows us to compute the catch per unit of effort (CUPE) for the different types of catch gear. Table 9.4 presents the data.

TABLE 9.4
Catch per Unit of Effort for Different Gear,
Ganga Section below Farakka

Gear/Net	1985–91		1992–5	
	Average	Range	Average	Range
Drift-gill	0.48	0.18–0.88	0.74	0.45–1.06
Purse seine	0.35	0.05–0.59	0.35	0.16–0.51
Trawl	1.49	1.12–1.94	1.82	1.42–2.01
Small seine	1.44	1.18–1.72	1.93	1.40–2.81
Bag	4.62	3.63–5.67	4.95	3.82–6.02
Set-barrier	1.93	1.31–2.56	1.62	1.45–2.01
Lift	1.17	0.71–2.46	0.34	0.24–0.59

Source: ICAR (1997).

Table 9.4 clearly indicates an increase in CUPE for all types of gear with the exception of lift and set barrier. These are not, however, the more important forms of gear, which are drift-gill, followed by trawl and seine. For all these three there has been a notable increase in catch per unit of effort. Some of this may be attributable to improvements in water quality, but account must also be taken of increased efficiency of some of the nets.

Total effort is established by looking at the total number of fishermen, amount of gear, number of boats, etc. Such data were collected only at intervals of 10 to 15 years, and no earlier data are available for the upper-middle stretch of the Ganga. For the lower stretch, a comparison can be made between 1982–3 and 1995–6. The main conclusion is that a large decrease in inventory has taken place. Two dominating forms of gear: drift-gill and dragnet declined by 26 per cent and 42 per cent respectively. Purse net use declined by 58 per cent. These declines can be attributed to a shift out of employment in this sector and into other sectors where wages and opportunities are better.

WATER QUALITY CHANGES

Water quality changes are summarized in Table 9.5 and Figures 9.3–9.5.[1] Substantial data are collected at each of the stations. Of these, the ones of interest to fisheries are: dissolved oxygen, pH, phosphate, and nitrate. Alkalinity can also be important, but is not included in Table 9.5 owing to space constraints. Instead, it is only commented on in the text below. The main conclusions on the trends in these substances are as follows.

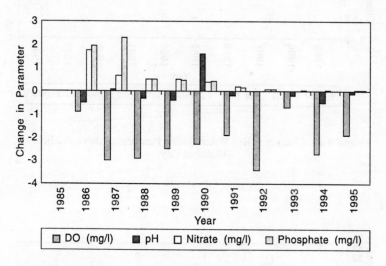

FIGURE 9.3:   Changes in Key Water Quality Parameters Above and Below Kanpur City

DISSOLVED OXYGEN (DO)

A significant change in dissolved oxygen was observed at almost all the centres in the middle stretch of the Ganga. The impact of the cities on water quality can be established by comparing the water quality above the city with that below the city and seeing how the difference changes over time. This shows that the DO measure of water quality continued to decline at Kanpur, Allahabad, and Varanasi up to 1992 and improved

[1]Figures 9.3 to 9.5 describe the change in water quality above and below the major cities of Kanpur, Allahabad, and Varanasi. Quality is measured in terms of DO, pH, nitrates, and phosphorous. No figures are presented for other locations due to space problems, but their results are described in the text.

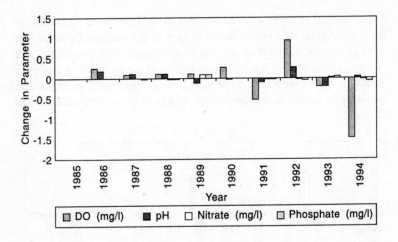

FIGURE 9.4: Changes in Key Water Quality Parameters Above and Below
Allahabad City

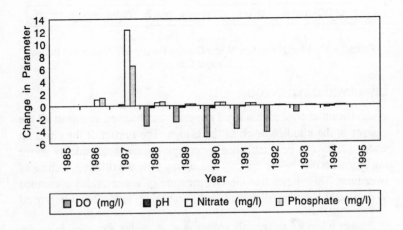

FIGURE 9.5: Changes in Key Water Quality Parameters Above and Below
Varanasi City

thereafter. In the lower stretch, DO at Nawabganj continued to decline, which explain the decline in fish catch there.

ACIDITY (pH)

A comparative study of fluctuations in pH value at two corresponding sampling points associated with each centre indicate that Kanpur and Varanasi occasionally found acidic water during 1988 and 1989 and during 1989 and 1990 respectively. The lower stretch was, however, conducive for aquatic life, with the water having the optimum pH.

ALKALINITY

Total alkalinity at Kanpur, Allahabad, and Varanasi in the middle stretch was significantly higher during 1986 and 1987 as shown in Table 9.5, indicating the impact of industrial pollution. However, during subsequent years, the content dropped significantly and almost attained the same level recorded at the above city level. The total alkalinity in the lower stretch varied between 56–174 ppm at Nabadwip, 58-160 ppm at Nawabganj, and 70–210 ppm at Uluberia. In other centres also the alkalinity remained within the range of 64–160 ppm, showing almost optimum level in the stretch below Farakka.

PHOSPHATE

The middle zone between Kanpur and Varanasi exhibited high phosphate content, particularly in the stretches of the Ganga below the cities (Figures 9.3 to 9.5). High phosphate levels at Kanpur and at Varanasi compared with the level above the city indicated aquatic pollution. This trend continued up to 1991; thereafter the phosphate levels came down, indicating considerable improvement of the Ganga above Farakka. In the lower stretch below Farakka, the phosphate level was low to medium, indicating a better environment.

NITRATE

Figures 9.3 to 9.5 present the data of nitrate level at the points above the city as well as below the city at three centres, namely Kanpur, Allahabad, and Varanasi, in the middle zone. An appraisal of the trends indicates pollution at all three centres up to 1990; thereafter, the situation improved considerably. The nitrate content was generally better in the lower stretch between Nawabganj and Nurpur.

CONCLUSIONS

This chapter has examined fisheries in the river Ganga. These have been an important resource and continue to be so. Ideally, the fish catch in the river and estimates of the change in that catch that could be attributed to improved water quality should be analysed. In practice, this was impossible to do, as the data required for such a sophisticated exercise were simply not available. Not only are the time series very short, they are incomplete for some stations and the corresponding data on effort are also incomplete. Hence, a partial quantitative analysis of changes in the fisheries situation had to be undertaken.

The total catch was looked at over the period 1985–93. There are few clear trends, but in a couple of cases where there was a trend decline, the trend appears to have been reversed (Sadiapur and Patna). In one location (Nawabganj) a positive time trend has become negative, and in one location (Konnagar) a negative time trend can be established. In all other cases no clear trends exist, although for one or two of them, (e.g. Lalgola) this may be due to lack of data.

The next set of data looked at were the trends in catch per unit of effort (CUPE) and total effort. For the former, there are encouraging signs that CUPE has risen for the major kinds of gear. Partly this reflects better equipment, but it may partly be due to increased fish productivity as a result of improved water. It is not possible to say how much is due to the one and how much to the other. Total effort has declined, mainly reflecting better employment opportunities in other sectors and is not a reflection on poorer fish productivity.

The last set of data analysed were those relating to water quality. Most of the key ones (dissolved oxygen, pH, alkalinity, phosphate, and nitrate levels) have shown a significant improvement since 1990 for most locations. This is, at least partly, the result of the GAP. It is also interesting to note that where the fish catch has declined sharply (for example Nawabganj), water quality indicators were also on the decline. Hence, we can conclude that the GAP has probably contributed positively to fish productivity, but it was not possible to quantify that impact further.

10

Agricultural Benefits of the Projects of the Ganga Action Plan

INTRODUCTION

Sewage treatment plants (STPs) under the Ganga Action Plan (GAP) provide agricultural benefits by supplying irrigation and non-conventional fertilizers. Partly treated water from STPs with significant amounts of nutrients is used by farmers for irrigation in the adjoining villages of the GAP projects. Also, the sludge generated in the treatment of household and industrial effluents by STPs was found to have significant fertilizer value when sludge from the GAP projects was analysed recently by the National Environmental Engineering Research Institute (NEERI, 1995). Therefore, STPs under the GAP, apart from reducing the pollution load of households and industrial effluents to meet pollution standards, provide irrigation and fertilizer benefits to farmers.

There are about 35 STPs under the GAP, with 13 located in Uttar Pradesh, 7 in Bihar, and 15 in West Bengal. Apart from providing irrigation, some of the STPs are supporting pisciculture, especially in West Bengal. All these STPs have the capacity to treat a wastewater volume of 919.82 million litres per day. Table 10.1 provides information about the wastewater treatment capacity, sludge generated, and revenue from sludge for the 35 GAP STPs. Table 10.5 gives influent and effluent characteristics of wastewater for the STPs.

IRRIGATION BENEFITS

The estimation of irrigation benefits from the GAP projects requires information about the quantity of partially treated water from STPs available for irrigation, the cropping pattern, and the per acre crop productivities of farms using the GAP water. Also, using information about the cropping pattern and per acre crop productivities for a control farm, a farm not using the GAP water, which is identical to the GAP farms in all other respects, incremental farm benefits using the GAP irrigation can be estimated. These are the difference between the net value of output produced by the GAP farms and the net value of output produced by the control farm. The control farm may be a farm either using other sources of irrigation, like canal or tubewell irrigation, or an unirrigated farm. Therefore, this must be taken into account when estimating the incremental irrigation benefits from the GAP projects.

Alternatively, incremental benefits from the GAP irrigation can be estimated using information about the cost of the GAP water and the cost of tubewell or canal water. However, it is difficult to estimate the cost of the GAP water because it is a by-product of the process of water pollution abatement.

To estimate the irrigation benefits of the GAP projects, a survey of 108 farmers in 12 villages around the Jajmau STPs at Kanpur and 116 farmers in 8 villages around the Dinapur STP at Varanasi was conducted in August, 1996. The survey collected information about the cropping pattern, crop productivities, input use, and sources of irrigation for the farms around the STPs, and general household characteristics like income, family size etc. of the farmers' households.

TABLE 10.1

Capacity of Wastewater, Sludge Generated, and Sludge Revenue of STPs

Name of STP	Capacity (MLD)	Sludge (MT/ Year)	Revenue from sludge (Rs. '000)
Uttar Pradesh			
1. Swarag Ashram, Rishikesh	0.32	25.71	0.71
2. Lakarghat, Rishikesh	6.00	482.14	13.34
3. Kankhal, Hardwar	18.00	144.6	40.01
4. Farrukhabad	2.70	216.96	6.00
5. Kanpur (Jajmau)	5.00	401.79	11.11
6. Kanpur (Jajmau)	36.00	2892.85	80.02
7. Chrome recovery pilot plant, Kanpur	0.004	0.32	0.01
8. Kanpur	130.00	10446.41	288.95
9. Allahabad	60.00	4821.42	133.36

Name of STP	Capacity (MLD)	Sludge (MT/ Year)	Revenue from sludge (Rs. '000)
10. Mirzapur	14.00	1125.00	31.12
11. BHU, Varanasi	10.00	803.57	22.23
12. DLW, Varanasi	12.00	964.28	26.67
13. Dinapur, Varanasi	80.00	6428.56	177.81
Bihar			
14. Chapra	23.60	1896.43	52.46
15. Patna Eastern Zone	4.00	321.43	8.89
16. Patna Saidpur	45.00	3616.07	100.02
17. Beur, Patna	35.00	2812.50	77.79
18. Patna Southern Zone	25.00	2008.93	55.57
19. Munger	35.00	2812.50	77.79
20. Bhagalpur	11.00	883.93	24.45
West Bengal			
21. Chandannagar	22.70	1824.10	50.45
22. Behrampur	4.00	321.43	8.89
23. Nabadwip	4.00	321.43	8.89
24. Kalyani	17.00	1366.07	37.79
25. Bhatpara Group E	10.00	803.57	22.23
26. Bhatpara Group B	23.00	1848.21	51.12
27. Titagarh	18.50	1486.60	41.12
28. Panihati	12.00	964.28	26.67
29. Baranagar-Kamarhati	40.00	3214.28	88.91
30. Garden Reach	47.00	3776.78	104.47
31. South Suburban (East)	30.00	2410.71	66.68
32. Howrah	45.00	3616.07	100.02
33. Serampore	19.00	1526.78	42.23
34. Bally	30.00	2410.71	66.68
35. Cossipore-Chitpore	45.00	3616.07	100.02
Total	*919.82*	*73914.30*	*2044.47*

In the case of the Jajmau STPs at Kanpur, the total household and tannery waste water intercepted is 148 MLD (million litres per day), of which only 41 MLD is currently treated. The remaining 107 MLD of waste water goes to fields untreated. A 15 kilometre canal carries treated and untreated waste water to irrigate 2000 hectares in 15 villages near Kanpur. The Dinapur STP at Varanasi has the capacity to treat 80 MLD. A 6 kilometre canal carries treated waste water to irrigate 1000 hectares in eight villages near Varanasi. In both the areas surveyed, there are three types of farmers: those using only the GAP water, those using both the GAP water and tubewell water, and those using only tubewell water for irrigation.

Tables 10.2 to 10.4 provide estimates of cropping patterns, crop yields, and the cost of chemical fertilizers based on the survey data of the two regions of Uttar Pradesh. There are significant differences in the cropping pattern, crop yields, and the fertilizer use between the farms using the GAP water and those which are not using the GAP water in both the regions surveyed. Differences with respect to these characteristics are also found between the farms using the GAP water only and those which are using both the GAP water and tubewell water. In the Varanasi area, farms using the GAP water, farms using both the GAP and tubewell water, and farms not using the GAP water have 30, 16, and 5 per cent of land under paddy respectively. However, there is no difference in the land used for wheat across different types of farms. A similar type of cropping pattern has been observed in the case of farms in the Kanpur area.

TABLE 10.2

Average Cropping Pattern in the Areas of Varanasi and Kanpur Under the GAP Projects (as percentage of Total Cropped Area)

Crops	Using GAP water	Not using GAP water
Paddy	28.50	5.00
Maize	2.75	5.75
Jowar	8.50	20.50
Pulses	1.25	7.50
Vegetables (I)	2.75	2.00
Wheat	40.00	39.50
Gram	1.00	3.00
Sarson	7.00	3.75
Pulses	0.50	5.25
Vegetables (II)	0.75	10.00
Others	8.00	8.00

Source: Survey data.

TABLE 10.3

Average Yield of Crops in the Areas of Varanasi and Kanpur Under the GAP Projects

Crops	Using GAP (quintals/ha)	Average Yield (Rs/ha)	Not using GAP (quintals/ha)	Average (Rs/ha)
Paddy	31.00	26,040	26.00	21,840
Maize	20.00	9000	16.50	7425
Jowar	10.25	4356	9.00	3825
Pulses	11.50	11,500	8.00	8000
Vegetables	–	–	–	–
Wheat	40.00	20,000	32.00	16,000
Gram	–	–	–	–
Sarson	8.50	9350	7.00	7875
Average returns		*16,837.00*		*9518.30*

Source: Survey data.

TABLE 10.4

Average Cost of Fertilizer in the Areas of Varanasi and Kanpur Under the GAP Projects

(Rupees per hectare)

Crops	Using GAP water	Not using GAP water
Paddy	155.00	800.00
Maize	125.00	315.00
Jowar	–	300.00
Pulses	–	–
Vegetables (I)	–	150.00
Wheat	150.00	725.00
Gram	–	–
Sarson	–	350.00
Vegetables (II)	–	150.00
Average Cost	*150.97*	*687.40*

Source: Survey data.

In both the areas surveyed, farms using the GAP water have higher crop yields or productivity in comparison to farms not using the GAP water. In the case of paddy and wheat, the irrigated crops, the GAP farms produce respectively 32 and 40 quintals per hectare in the Varanasi area, and 30 and 40 quintals per hectare in the Kanpur area. The corresponding yields for the non-GAP farms are 26 and 32 quintals per hectare in the Varanasi area. The survey data shows a substantial difference in the use of conventional fertilizers between the GAP and the non-GAP farms. In the Varanasi area, the GAP farms spent Rs 150 per hectare for paddy and wheat while the corresponding amounts spent by the non-GAP farms are Rs 800 and Rs 850 per hectare. In the Kanpur area, the GAP farms spent Rs 150 per hectare for wheat while the corresponding amount spent by the non-GAP farms was Rs 600 per hectare.

The third category of farms—those using both GAP water and irrigation from other sources, especially tubewell irrigation—are found to have higher crop yields and higher fertilizer use in comparison to the GAP farms.

The 35 STPs under the GAP have a waste water capacity of 919.82 MLD as given in Table 10.1. Given the volume of waste water and an estimate of waste water capacity per day required to irrigate one hectare of land, the annual irrigation potential of STPs can be estimated. Data obtained from STPs in the Varanasi and Kanpur areas show that the Dinapur STP in Varanasi has a waste water capacity of 80 MLD, irrigating 1000 hectares of land, and STPs in the Kanpur area have a waste water capacity of 143 MLD, irrigating 2000 hectares. An estimate based on this data shows that to irrigate 1000 hectares of land, a STP of 74.3 MLD capacity is needed. On the basis of this estimate, it is calculated that the 919.82 MLD waste water capacity of STPs of GAP can irrigate 12,380 hectares of land.

The estimates of average annual yields per hectare of cropped area based on the survey data for GAP and non-GAP farms are given as Rs 16,837 and Rs 9518 respectively. Therefore, the incremental benefits per hectare of irrigated land in GAP farms is estimated as Rs 7319. The annual incremental benefits from irrigating 12,380 hectares of land by the GAP projects can now be estimated at Rs 90.61 million at 1995–6 prices.[1]

In the case of irrigated crops like paddy and wheat, the estimated values of conventional fertilizers used are given as Rs 155 and Rs 150

---

[1] A reviewer has questioned the valuation of the benefits in terms of gross value of returns rather than the net values after deduction of expenses. Taking the gross values is correct when the change in the environmental variable is small and when allowance is made for the fact that farmers can optimize with respect to crop mix, inputs, etc. In that case, it can be shown that the benefits of an improvement in an external factor such as water quality is the gross increase in yield (EC 1995: Part II, chapter 7).

for GAP farms, and Rs 800 and Rs 725 for non-GAP farms. The estimated costs of conventional fertilizers per hectare of the GAP and non-GAP farms based on the survey data are respectively given as Rs 150 and Rs 687. Therefore, the per hectare incremental benefit for savings in fertilizer cost due to the GAP projects is estimated as Rs 535. The annual incremental benefits from the savings in the fertilizer cost can now be estimated as Rs 6.6 million at 1995–6 prices.

BENEFITS FROM USING THE SLUDGE

The sludge generated in the process of treating wastewater by STPs has been found to have fertilizer potential. The data for wastewater volume in terms of MLD and the annual quantity of sludge generated for STPs at Mirzapur and Kankhal at Hardwar shows that for a wastewater volume of one MLD, about 8.036 metre of sludge can be generated annually. Given the total wastewater volume of 919.82 MLD from all STPs of the GAP, the total amount of sludge generated annually by the GAP projects can be estimated as 73,914.30 mt (Table 10.1). NEERI (1995) provides the chemical analysis of sludge from some STPs from the Varanasi and Kanpur areas as given in Table 10.6. According to the NEERI analysis, the average concentration of fertilizers in the sludge is estimated as 18.37 kg/mt, 5.08 kg/mt, and 4.68 kg/mt respectively for nitrogen, phosphorous, and potassium (Table 10.6). Given these estimates, annual fertilizer potential of sludge from the GAP projects can be estimated as 1357.805 mt of nitrogen, 375.485 mt of phosphorous, and 345.919 mt of potassium.

TABLE 10.5

Wastewater Capacity and Characteristics of the GAP Sewage Treatment Plants

Name of STP	Capacity (MLD)	BOD (mg/l)		TSS (mg/l)	
		Influent	Effluent	Influent	Effluent
Uttar Pradesh					
1. Swarag Ashram,Rishikesh	0.32	125	48	471	175
2. Lakarghat, Rishikesh	6.00	140	42	675	125
3. Kankhal, Hardwar	18.00	106	14	360	20
4. Farrukhabad	2.70	158	26	380	185
5. Kanpur(Jajmau) 1	5.00	261	30	679	121
6. Kanpur(Jajmau) 2	36.00	586	206	1400	482

Name of STP	Capacity (MLD)	BOD (mg/l)		TSS (mg/l)	
		Influent	Effluent	Influent	Effluent
7. Chrome recovery pilot plant	0.004	–	–	–	–
8. Kanpur	130.00	Under construction			
9. Allahabad	60.00	Under construction			
10. Mirzapur	14.00	135	17	267	32
11. BHU, Varanasi	10.00	112	16	97	21
12. DLW, Varanasi	12.00	66	9	184	28
13. Dinapur, Varanasi	80.00	234	42	390	62
Bihar					
14. Chapra	23.60	Under construction			
15. Patna Eastern Zone	4.00	Under construction			
16. Patna Saidpur	45.00	120	109	216	44
17. Beur, Patna	35.00	82	21	84	13
18. Patna Southern Zone	25.00	Under stabilization			
19. Munger	35.00	Under construction			
20. Bhagalpur	11.00	Low flows reaching the STP			
West Bengal					
21. Chandan Nagar	22.70	65		25	
22. Behrampur	4.00	–	–	–	–
23. Nabadwip	4.00	–	–	–	–
24. Kalyani	17.00	–	–	–	–
25. Bhatpara Group E	10.00	–	–	–	–
26. Bhatpara Group B	23.00	152	66	–	–
27. Titagarh	18.50	167	41	–	–
28. Panihati	12.00	–	–	–	–
29. Baranagar-Kamarhati	40.00	–	–	–	–
30. Garden Reach	47.00	Under construction			
31. South Suburban (East)	30.00	Under construction			
32. Howrah	45.00	115	31	–	–
33. Serampore	19.00	168	27	–	–
34. Bally	30.00	–	–	–	–
35. Cossipore-Chitpore	45.00	Under construction			

The retail prices of fertilizers, nitrogen, diammonium phosphate, (DAP), and potash are respectively given as Rs 4000, Rs 12000, and Rs 6000 per mt during the year 1995–6 in the surveyed areas. Given these prices, the value of fertilizers in the total sludge is estimated as Rs 5.43

TABLE 10.6
Characteristics of Sludge from Sewage Treatment Plants

District	Varanasi						Kanpur			
Name of STP	BHU		Dinapur		DLW		Jajmau 1		Jajmau 2	
Type of sludge	Digested	Dried	Digested	Dried	Digested	Dried	Digested	Dried	Digested	Dried
pH	6.70	6.30	6.90	6.40	6.80	6.30	8.90	7.70	8.60	8.00
Temperature C	31	30	32	31	31	33	29	23	29	24
Total organic carbon	183	104	100	97	88	45	127	117	140	160
Total nitrogen (N)	17.70	14.20	10.70	17.40	18.60	26.70	17.50	19.00	20.20	21.78
Nitrate nitrogen (NO_3)	BDL	BDL	BDL	BDL	BDL	BDL	BDL	BDL	BDL	BDL
Total phosphorus (PO_4)	5.00	5.40	5.40	6.00	5.60	6.60	4.30	3.20	5.40	3.90
Potassium	4.20	3.80	6.30	4.60	4.50	3.00	8.00	4.20	5.00	3.20
Boron	0.002	0.002	0.001	0.001	0.001	0.001	0.004	0.005	0.005	0.005
Cadmium	0.003	0.001	0.330	0.250	BDL	trace	0.045	0.049	0.066	0.074
Chromium	0.60	3.00	0.20	0.20	0.60	2.00	27.50	15.30	21.40	11.50
Copper	0.40	0.50	1.50	1.80	0.20	0.20	0.40	0.30	0.40	0.20
Iron	13.60	13.20	18.80	17.00	16.80	15.10	11.40	12.50	10.00	13.10
Manganese	0.20	0.20	0.30	0.30	0.20	0.20	0.10	0.20	0.20	0.20
Nickel	0.10	0.10	0.50	0.60	0.30	0.70	0.04	0.03	0.04	0.01
Lead	0.05	0.09	0.11	0.16	0.60	0.70	0.25	0.12	0.19	0.16
Zinc	1.40	4.80	3.80	9.10	1.70	4.00	1.70	1.90	1.40	1.30

| District | Varanasi | | | | | | Kanpur | | | |
| Name of STP | BHU | | Dinapur | | DLW | | Jajmau 1 | | Jajmau 2 | |
Type of sludge	Digested	Dried	Digested	Dried	Digested	Dried	Digested	Dried	Digested	Dried
Pesticides (µg/kg dry wt.)										
—DDT	BDL	BDL	2699	800	BDL	BDL	1137	639	392	351
—BHC	1290	95	96	32	171	8	922	457	733	320
—Aldrin	BDL	BDL	BDL	BDL	BDL	BDL	BDL	BDL	BDL	BDL
—Dieldrin	BDL	BDL	BDL	BDL	BDL	BDL	BDL	BDL	BDL	BDL
—Endosulphan	BDL	BDL	BDL	BDL	BDL	BDL	1385	828	437	409
Bacterial count (CFUx105/g)										
—Total coliform	593	93	1231	262	1034	150	79,000	9500	53,000	8500
—Fecal coliform	298	47	40	153	600	120	11,000	2100	26,000	5300
—Fecal streptococci	4	13	188	72	–	12	34,000	4200	180,000	11,000

Notes: 1. All values except temperature, pH, and bacterial count are expressed on dry wt. basis in g/kg.
2. BDL: Below detectable limit
3. CFU: Colony forming units.
4. * Calculated on dry weight basis
5. BHU: Banaras Hindu University
6. DLW: Diesel locomotive works;
7. TWT: Tannery waste treatment
8. Jajmau 1: Sewage treatment plant
9. Jajmau 2: Tannery + municipal waste treatment plant.

Source: National Environmental Engineering Research Institute (1995), *Annual Report, 1995*.

million for nitrogen, Rs 4.51 million for phosphorous, and Rs 2.08 million for Potash. Therefore, the total value of fertilizers in the sludge annually generated by the GAP projects is estimated as Rs 12.02 million at 1995–6 prices (Table 10.7).

TABLE 10.7

Estimated Value of Fertilizers in Total Sludge Generated by STPs under the GAP Projects

Name of the fertilizer	Concentration of fertilizer in sludge (kg/mt)	Amount of fertilizer in total sludge (kg)	Fertilizer Value in total sludge (Rs million)
Nitrogen (N)	18.37	1,357,806	5.4
Phosphorus (PO₄)	5.08	375,485	4.5
Potassium	4.68	345,919	2.1
Total		*2,079,210*	*12.0*

CONCLUSIONS

The agricultural benefits of the GAP come in three forms. First, there are the irrigation benefits arising from the partially treated water that is released to farms in the GAP area. These were estimated from a survey comparing the cropping patterns, input use, sources of irrigation, and yields for farms receiving the GAP water with similar farms not receiving the GAP water. The part of higher yields due to the GAP water was estimated at Rs 7319/hectare/year, for 12,380 hectares, which amounts to Rs 90.61 million anually, at 1995–6 prices.

The second benefit is from the fertilizer value of the irrigation water. It is observed that the GAP farms apply less fertilizer than the non-GAP farms. The incremental benefits of this function of the GAP water is Rs 535/hectare/year. Aggregating across all the GAP-affected farmland gives a total benefit of Rs 6.62 million.

Finally, there are the benefits of the use of sludge provided by the STPs. About 74,000 MT of sludge are generated by the 35 STPs under the GAP. These are valued in terms of the market prices of their nutrient content (nitrogen, phosphorous, and potassium) at Rs 12.02 million a year.

These benefits add up to Rs 109.26 million per year. This figure has been taken in the net benefit calculations in Chapter 12.

11

Biodiversity of the Ganges: Impacts of the Ganga Action Plan

INTRODUCTION

The Ganga river, like many rivers of the world, sustains a diverse flora and fauna, which has not only helped the river in maintaining the pristine purity of its water, but has served as a resource for humankind since time immemorial. The total species of the world recorded so far are about 1.5 million (Fitter, 1986), out of which about 127 thousand are from India. The Ganga river at present sustains approximately 2500 species or more, from micro-organisms to mammals. Though some reference has been made to certain biota of the river in mythological books, there has been no systematic record made of the biodiversity of the Ganga. As a consequence, very little time series scientific data on the subject is available.

We are grateful to the National River Conservation Directorate of the Government of India, New Delhi, for sponsoring a small project to compile and analyse the data available on the biodiversity of the Ganga River, which was used in this cost–benefit analysis of the Ganga Action Plan. This work was undertaken by Dr P. Sinha of Patna University. The following section deals with species of international conservation significance found in the Ganga river and the subsequent sections with species (other than those in Fishery statistics) in the Ganga river or on the banks that have significant commercial use value and with overall assessment of 'ecosystem health' not captured elsewhere.

SPECIES OF INTERNATIONAL CONSERVATION SIGNIFICANCE FOUND IN THE GANGA RIVER

MAMMALS

Ganga River Dolphin

The Ganga River Dolphin (*Platanista gangetica*) is listed as endangered by the International Union for Conservation of Nature (IUCN). No time series data on its population status is available. However, about two decades ago, their number was estimated to be 4000 to 5000 (Jones, 1982) and the current estimate is 2000 to 2500 in its entire distribution range (Sinha, 1996). In the last century it was distributed in the Ganga, from the foothills of the Himalayas (Hardwar), the place up to which the river was navigable, to the tidal zone of the river in the Bay of Bengal. Currently, it is distributed between Bijnor (approximately 125 kilometres downstream of Hardwar) and Kakdwip near Ganga Sagar Island at the mouth of the Ganga in the Bay of Bengal.

Threats to the distribution of Cetaceans (the family to which the Ganga River Dolphin belongs) are more directly related to prey distribution (Leatherwood and Walker, 1979; Whitehead and Carscadden, 1985). Extensive and intensive juvenile fishing in the Ganga is reducing availability of food to this dolphin and, thus, posing a threat, as it feeds mainly on juvenile and other smaller varieties of fish.

The major threats to this animal are over-exploitation and habitat destruction. In the past, the Ganga River Dolphin was killed, both intentionally—mainly for oil which is used as fish bait and liniment—and incidentally, as a consequence of other exploitation schemes. The practice of directed killing was common even in the last decade. No intentional killing has been noticed in the last 5 to 6 years. Now the exploitation is mainly due to incidental catching of the dolphin in monofilament nylon gill-nets. The Ganga River Dolphin is a blind animal and its power of sight has been substituted for by well developed powers of echolocation. The animal fails to receive echolocation from monofilament gill-nets, which are made of very fine nylon threads.

Habitat destruction is another major threat. This has mainly been due to the construction of dams and barrages, as well as various Water Development Projects. Within the last four decades, barrages have been constructed on all the tributaries, as well as the main stem of the Ganga, which have isolated the dolphin population into several small sub-populations, making it genetically more vulnerable. Such obstructions have also adversely affected the migration of species on which the dolphin preys. The extraction of river water for irrigation and other purposes, together with the heavy rate of siltation in the river, is also degrading the dolphin's habitat.

Pollution sources, both point and non-point, are also affecting the river ecosystem. The Ganges basin receives 1.15 million metric tonnes of chemical fertilizers of all kinds annually (Das Gupta, 1984). It has also been estimated that 2573 tonnes of pesticides end up in the Ganges river system every year. Recent studies show high levels of organochlorines, for example DDTs, HCHs, HCB, aldrin, dieldrin, PCBs, and heavy metals, for example Fe, Mn, Zn, Cu, Pb, Ni and Cd, butylins in the tissues of the Ganga River Dolphin and fish collected from the stomach of the dolphin (Kannan *et al.*, 1993, 1994, 1997). The mean DDT concentration was 0.01 milligram per litre in Ganga water, as reported by ITRC Lucknow during 1985–91, whilst the average bioconcentration factors in the blubber of male and female dolphins collected from the Ganga at Patna were 6.9×10^5 and 1.2×10^6 respectively (Kannan *et al.*, 1994). Though no direct evidence of deaths of Ganga River Dolphin due to toxic chemicals is available, these chemicals might be affecting the various vital organs, as well as processes, and further investigation is needed.

The dolphin was given protection even in ancient India, by the emperor Ashoka the Great in the third century BC. He passed a government decree to extend legal protection to various animals, including the Ganga River Dolphin.

The Ganga River Dolphin has been given legal protection under the Wildlife (Protection) Act, 1972. It is one of the Schedule I animals of the Indian Wildlife (Protection) Act being given full protection, and is included in Appendix I of CITES. However, the efficacy of the legal measures is not satisfactory. A 50 kilometre stretch of the Ganga, between Sultanganj and Kahalgaon in Bihar, was turned into the Vikramshila Gangetic Dolphin Sanctuary in 1991. This is the only protected area for this animal.

International organizations like the Cetacean Specialist Group of the IUCN's Species Survival Commission are giving top priority to the conservation of this species. The Asian River Dolphin Committee, under the aegis of the Cetacean Specialist Group, is making a special effort to save this animal from extinction throughout its distribution range in India, Nepal, and Bangladesh.

THE IMPACT OF THE GANGA ACTION PLAN (GAP): For the first time a Dolphin Conservation Project was launched in 1991 under the GAP to evaluate the current status of the Ganga River Dolphin. Detailed study was undertaken on various aspects of biology, behaviour, and ecology of the animal. Such study is essential for long-term conservation measures. Various short term conservation measures were also initiated. A mass awareness campaign was a major component of the conservation efforts. Mass media, both national and international, gave wide publicity and strong support to this programme. The recent release of a live dolphin back into the Ganga and protest by the fishermen themselves against killing of the dolphin in the Ganga in April 1997 at Patna are encouraging results of the conservation effort under the Dolphin Conservation Project of the GAP. Under the project, alternatives to dolphin oil as fish bait are being tried and very encouraging results have been obtained. Experiments are still being carried out to provide a better scientific database.

The pollution abatement schemes under the Ganga Action Plan will definitely improve the habitat of the dolphin in the Ganga and its tributaries. It will not only improve the water quality of the river, but also the ecology of the river as a whole, and will definitely have a direct bearing on the fisheries of the Ganga. All these impacts will help dolphin conservation in the Ganga.

The pollution abatement schemes will also reduce the load of toxic chemicals such as heavy metals and pesticides in the river and, in turn, the impact on fishes and dolphins.

Irrawady Dolphin and Finless Porpoise

Both the Irrawady Dolphin (*Oraella brevirostris*) and the Finless Porpoise (*Neophocaena phocaenoides*) were categorized until recently as threatened, although they are not included in the latest IUCN Red List (Baillie J. and B. Groombridge (eds), 1996). Their estimated populations are not known. Their distribution within the Ganga is reported from the Sundarbans area of the deltaic zone. Occasionally, incidental killing has been reported from the Sundarbans area and also from Chilka Lake in Orissa.

No specific conservation measures for either species have been initiated in India. Both are listed as a Schedule II animals in the Indian Wildlife (Protection) Act and are included in Appendix II of CITES.

THE IMPACT OF THE GANGA ACTION PLAN (GAP): The GAP will probably have no direct impact on these species as they are found only in the brackish water at the river mouth. Nevertheless, pollution abatement schemes will have some indirect bearing on the species.

Common Otter

The Common Otter (*Lutra lutra*) was listed as vulnerable by the IUCN, although it was not included in the 1996 Red List. Its population is unknown, but it is found in the Ganga and its tributaries in Uttar Pradesh, and in the Hooghli River in West Bengal. The main threats are habitat degradation due to intensive agriculture, industrialization, expanding cities, and increasing human disturbance. Forest destruction in general, and riparian vegetation in particular, are serious conservation problems. Hunting pressures come from two sources: those seeking certain parts of the body, which in some regions are believed to have medicinal properties, and those seeking the luxurious pelts. CITES parties have recorded 712 skins and 52 live specimens between 1979 and 1981 (Foster-Turley *et al.*, 1990).

Specific conservation efforts have been initiated by the Wildlife Institute of India to save the otters. The animal is listed in Appendix I of CITES (which prohibits international trade in the species) and as a Schedule II animal under the Indian Wildlife (Protection) Act, 1972.

THE IMPACT OF THE GANGA ACTION PLAN (GAP): Pollution abatement schemes under the GAP will improve fish availability in the river, that is the food of the otters. Though no information is available on the toxic load in otters, the GAP will definitely reduce the load of hazardous chemicals in their food chain.

Smooth-coated Indian Otter

The Smooth-coated Indian Otter (*Lutra perspicillata*) is categorized by the IUCN as vulnerable, though there is insufficient information available. Its population is not known, but it can be found throughout the stretch of the Ganga and its tributaries from the foothills of the Himalayas to the Sundarbans. The main threats are habitat degradation due to intensive agriculture, industrialization, expanding cities, and increasing human disturbance. Forest destruction in general, and that of riparian vegetation in particular, is a serious conservation problem. It is also threatened by hunting pressures from two sources: those seeking certain parts of the body for medicinal properties and those seeking the pelts. It is listed in Appendix II of CITES and Schedule II of the Indian Wildlife (Protection) Act, 1972. Specific conservation measures have been initiated by the Wildlife Institute of India.

THE IMPACT OF THE GANGA ACTION PLAN (GAP): Pollution abatement schemes under the GAP will improve fish availability in the river, which is the food of otters. Though no information is available on the toxic load in these otters, the GAP will reduce the load of hazardous chemicals in their food chain.

Asian Small-Clawed Otter

The Asian Small-clawed otter (*Aonyx cynerea*) was categorized as threatened, but is not on the 1996 IUCN Red List. There is sufficient information about its population. It is found throughout the Ganga and its tributaries, from the foothills of the Himalayas to the Sundarbans. The main threats are habitat degradation due to intensive agriculture, industrialization, expanding cities, and increasing human disturbance. It is also threatened by hunting. Forest destruction in general, and riparian vegetation in particular, are serious conservation problems. Hunting pressures are as with the otters previously mentioned. The animal is listed in Appendix II of CITES and Schedule I of the Wildlife (Protection) Act, 1972.

THE IMPACT OF THE GANGA ACTION PLAN (GAP): Pollution abatement schemes under GAP will improve the availability of fish in the river, which is the food of otters. As in the case of the other otters, though no information is available, reducing the toxic load of the river water will reduce the toxic load in the otters and their food chain.

BIRDS

Large Whistling Teal

The Large Whistling Teal (*Dendrocygna bicolor*) is classified by the Zoological Survey of India as endangered. It is found on the middle and lower Ganga, but its population is unknown. In recent years it has become scarce as a result of the shrinkage of its habitat due to enhanced encroachment by man. There are no specific conservation measures, but it is included in Schedule III of the Wildlife (Protection) Act, 1972.

THE IMPACT OF THE GANGA ACTION PLAN (GAP): As it feeds on small fish, the GAP should have some indirect impact on this species.

Osprey

The Osprey (*Pandion haliaetus*) is also classified by the Zoological Survey of India as endangered. Its population is unknown, but it is found on the middle and lower Ganga. The Osprey is rather scarce in India. In many localities, young birds do not survive to the fledgling stage as a result of feeding on fish in pesticide contaminated water. No specific conservation measures have been initiated for this species, but it is under Schedule III of the Wildlife (Protection) Act, 1972.

THE IMPACT OF THE GANGA ACTION PLAN (GAP): With a reduction in the amount of pesticides in the river due to various pollution abatement schemes under the GAP, the species should be helped by the improvement in its food chain chemical load.

Ring-tailed or Pallas' Fishing Eagle

The Ring-tailed or Pallas' Fishing Eagle (*Haliaeetus leucoryphus*) is categorized by the IUCN as vulnerable. It is found in the middle and lower reaches of the Ganga, but its population is unknown. Its usual prey is fish, frogs, turtles, wild fowls, etc. Pesticide contamination of its food chain is a threat to this species. However, there are no specific conservation measures.

THE IMPACT OF THE GANGA ACTION PLAN (GAP): Reduction in the river water pollution should improve the quality of the eagle's food chain.

Indian Skimmer

The Indian Skimmer (*Rynchops albicollis*) is assigned an endangered category by the Zoological Survey of India and is also found in the middle and lower reaches of the Ganga. It feeds chiefly on fish, and though the exact cause of the decline in population of this animal is not clearly understood, heavy pollution of the larger Indian rivers might possibly have played some role in the dwindling of its numbers.

THE IMPACT OF THE GANGA ACTION PLAN (GAP): There are no specific conservation measures, but reduction of the pollution load in the Ganga and its tributaries as a result of the GAP may help restoration of the species.

REPTILES

Gharial

The Gharial (*Gavialis gangeticus*) is categorized by the IUCN as endangered. In the nineteenth century the Gharial was in abundance in many areas, but is now extinct or extremely depleted throughout its former range. In 1970, the estimated Indian population was not more than 75 animals (Sharma, 1991). In 1976, the Gharial population rose to 107 (Rao *et al.*, 1995). The Zoological Survey of India in its 'Faunal Resources of the Ganga' gave an approximate number of 141 Gharials, including juveniles, in the major rivers of Uttar Pradesh, Rajasthan, and Madhya Pradesh, less than 10 each in Bihar and Orissa and around 10 in Assam (Sharma, 1991). During the Population and Habitat Viability Assessment Workshop of the Gharial held at Gwalior in January 1995, it was reported by the working group that approximately 1200 to 1300 Gharials are surviving in the National Chambal Sanctuary, less than 100 animals each in the Ghaghara and Gandak rivers, and less than 15 animals in the Ganga (Rao *et al.*, 1995).

The Gharial can be found throughout the river's stretch, from the foothills of the Himalayas to the tidal zone. It is also distributed in all the major tributaries of the Ganga, including the Brahmaputra in India, Nepal, and Bangladesh.

Principal threats are habitat destruction and limited distribution. By the end of the 1960s, the Gharial had dwindled to a trace of its former abundance due to the combined effect of shooting and hunting for skin and meat, and loss of habitat in the riverine system. Although this has not been extensively documented, the indiscriminate killing of Gharial, either directly or accidentally, in modern fishing equipment such as nylon gill-nets, has been largely responsible for the extermination of the species from river stretches formerly known to abound with these animals.

Some of the other direct threats are the removal of eggs, dynamite fishing, bamboo rafting and subsidized predation. The indirect threats are habitat degradation due to the construction of dams and barrages, pollution, riparian cultivation, that is farming in riparian zones, construction of embankments, and other such Water Development Projects.

A Gharial rehabilitation programme was launched in the mid-1970s as an initiative of both the Government of India and international conservation bodies. The animal is included under Schedule I of the Wildlife (Protection) Act, 1972 and classified under Appendix I of CITES. Many captive breeding centres were established. The juveniles, after rearing, are released into natural habitats. Many Protected Areas. have been established to protect the Gharial.

THE IMPACT OF THE GANGA ACTION PLAN (GAP): The Ganga Action

Plan does not have any specific component for the conservation of Gharials. However, research undertaken in the upper reaches between the Haridwar and Kanpur stretch of the Ganga, using crocodiles and turtles as indicator species, has helped in collecting information on the current status of the species in this stretch. Moreover, the pollution abatement schemes of the GAP will definitely improve the habitat and food availability of the animal.

Mugger or Marsh Crocodile

The Mugger or Marsh Crocodile (*Crocodylus palustris*) is categorized by the IUCN as vulnerable. Its present population is estimated to be around 5000 in India (Sharma, 1991) with a distribution reported from many parts of the Ganga and its tributaries including Gandak and Ghaghara in India, and Narayani and Karnali in Nepal. It was sighted at Brijghat, Narora, and Ramghat in the Ganga in 1993–4 (Rao, 1996).

Principal threats to the crocodile are habitat destruction and illegal hunting. While illegal skin trade was a major problem in the 1950s and 1960s, the current threats to the mugger crocodile are principally drowning in fish nets, egg predation by people, habitat destruction, and use of crocodile parts for medicinal purposes (Groombridge, 1982).

In India, the main threats to the survival of the Mugger are set net fishing (nylon gill-nets); egg predation for food by humans, mongooses, jackals, monitor lizards, and sloth bears; hatchling predation by birds like herons and storks; natural calamities like flooding and desiccation; habitat destruction by the construction of dams and barrages and other water development projects; and the medicinal use of crocodile parts and their eggs by human beings.

As a conservation measure, a large-scale captive rearing programme was initiated in 1975. The project has collected eggs from the wild, as well as produced young in captivity from captive adult breeding stock. the resulting juveniles have been used to restock natural populations in 28 National Parks, Wildlife Reserves, and Crocodile Sanctuaries throughout the country. Restocking efforts have declined in recent years, in part due to lack of suitable release sites. The Mugger is included in Schedule I of the Wildlife (Protection) Act, 1972, and Appendix I of CITES.

THE IMPACT OF THE GANGA ACTION PLAN (GAP): The aforementioned research undertaken in the upper-middle reaches between the

Haridwar and Kanpur stretch of the Ganga using crocodiles and turtles as indicator species has helped in collecting information on the current status of the species in this stretch. The pollution abatement schemes of the GAP will definitely improve the habitat and food availability of this animal.

Saltwater Crocodile

The Saltwater Crocodile (*Crocodylus porosus*) was also categorized by the IUCN as vulnerable, although it is not on the 1996 Red List. The total Indian population of this crocodile may be 170 to 330 (Sharma, 1991). It is found in the lower reaches of the river, that is in the deltaic zone in the Sundarbans. Its principal threats are illegal hunting and habitat destruction. Its hides are in high demand due to their large size. Thus, the main reason of its over-exploitation is organized killing. The habitat degradation is mainly due to loss of mangrove cover and increasing human activities in the estuarine zone.

This crocodile is a Schedule I animal of the Wildlife (Protection) Act, 1972, and is included in Appendix I of CITES. Several captive breeding centres in Andhra Pradesh, Tamil Nadu, Orissa, and West Bengal are successfully rearing this species, which breeds rapidly in captivity. Breeding and recruitment take place principally in rivers with significant freshwater input or in freshwater swamps. A restocking programme in the Bhitarkanika National park in Orissa has been quite successful, with over 1000 crocodiles being released prior to 1989 (Messel *et al.*, 1992).

THE IMPACT OF THE GANGA ACTION PLAN (GAP): It is probable that GAP will not have much impact on this species. However, some positive impact on water quality and fish availability can be expected in the estuarine zone of the Sundarbans.

Ganga Soft-shelled Turtle or Indian Soft-shelled Turtle

The Ganga Soft-shelled Turtle or Indian Soft-shelled Turtle (*Aspideretes gangeticus*) is considered by the Zoological Survey of India as an endangered species. Exact data on its population are unknown, but their number has reduced drastically in the recent past (Sharma 1991). It is distributed in the Ganga and its major tributaries in India. Threats to the species come from habitat destruction due to crop farming in the riparian zones where they lay eggs, loss of vegetation cover in the riparian zones, construction of dams and barrages proving barriers in migration to their breeding grounds, excessive killing of adults for flesh, and over-exploitation of eggs for food.

The species has been given legal protection and is included under Schedule I of the Indian Wildlife (Protection) Act, 1972 and Appendix I of CITES. The stretch of the Ganga near Varanasi has been declared a Turtle Sanctuary.

THE IMPACT OF THE GANGA ACTION PLAN (GAP): A captive breeding and rearing programme to restock the natural population in the Ganga was undertaken under the Ganga Action Plan by the ,Uttar Pradesh Wildlife Department. The pollution abatement schemes under GAP will help in improving the water quality and habitat of the animal.

Peacock-marked Soft-shelled Turtle

The Peacock-marked Soft-shelled Turtle (*Aspideretes hurum*) is considered by the Zoological Survey of India to be endangered. Again, the exact population is unknown, but the number has declined drastically. It is found in the lower reaches of the Ganga river. Threats to the species are the same as those for the Ganga Soft-shelled Turtle. It is categorized under Schedule I of the Wildlife (Protection) Act, 1972 and in Appendix I of CITES and is nominally protected by legislation, but measures for conservation are not followed strictly.

THE IMPACT OF THE GANGA ACTION PLAN (GAP): Pollution abatement schemes will help in improving habitat conditions in the river.

North Indian Flap-shelled Turtle

The North Indian Flap-shelled Turtle (*Lissemys punctata*) is not categorized by the IUCN, but it is threatened according to the Zoological Survey of India. Again, although exact numbers are not known, the population has been drastically reduced in recent years. It is distributed throughout the Ganga river system. The species is threatened by over-exploitation of adults and their eggs for protein rich food, and also due to habitat destruction. The nests of these turtles are often raided by predators like man, otters, mongooses, jackals, and dogs. Hydroelectric dams and barrages have also proved to be barriers in their migration to breeding grounds. The turtle is included under Schedule I of the Indian Wildlife (Protection) Act, 1972 and in Appendix I of CITES. Export of the adults and their products is prohibited.

THE IMPACT OF THE GANGA ACTION PLAN (GAP): A captive breeding and rearing programme to restock the natural population in the Ganga was undertaken under the Ganga Action Plan by the Uttar Pradesh Wildlife Department. The pollution abatement schemes under GAP will help in improving the water quality and habitat of the animal.

Narrow-headed Soft-shelled Turtle

The Narrow-headed Soft-shelled Turtle (*Chitra indica*) is considered by the Zoological Survey of India to be possibly threatened, as its population has become considerably reduced. It is listed as vulnerable by the IUCN. It is distributed in the Ganga and its tributaries in Uttar Pradesh, Bihar, and West Bengal. Threats to the species come from excessive killing of adults and over-exploitation of eggs for food, habitat destruction due to the construction of dams and barrages, and also riparian cultivation, which destroys the breeding areas. It is categorized as a Schedule IV animal under the Indian Wildlife (Protection) Act, 1972, and is nominally protected by legislation.

THE IMPACT OF THE GANGA ACTION PLAN (GAP): Improvement in the water quality and reduction of the pollution load will in turn help to improve the habitat of the animal.

North Indian Freshwater Tortoise or Black-spotted Pond Turtle

The North Indian Freshwater Tortoise or Black-spotted Pond Turtle (*Geochlemys hamiltoni*) is categorized as vulnerable by the Zoological Survey of India. Its population is not known, but it is found throughout the Ganga and its tributaries from Uttar Pradesh to West Bengal. Threats to the species come from over-exploitation of eggs and adult individuals for food, habitat destruction due to the commercial removal of sand from nesting areas, and the construction of dams/barrages which obstruct the movement of the species to its nesting sites. No specific conservation measures have been initiated and legal protection is nominal, but it is included under Schedule I of the Indian Wildlife (Protection) Act, 1972.

THE IMPACT OF THE GANGA ACTION PLAN (GAP): The pollution abatement schemes will have a positive impact on the river water and ultimately on the habitat of the animal.

Brahminy River Turtle or Crowned River Turtle

The Brahminy River Turtle or Crowned River Turtle (*Hardella thurjii*) is considered to be extremely vulnerable by the Zoological Survey of India. Its population is unknown, but it is found throughout the Ganga and its tributaries, including the Brahmaputra. Over-exploitation of adults, especially in the eastern part of India and Bangladesh, poses a threat to the species. Eggs are exploited by animal predators and humans for food, and habitat destruction due to various anthropogenic activities also poses serious threats to the species. It is included under Schedule II of the Wildlife (Protection) Act, 1972. Generally, the species is given legislative protection throughout its range, but the conservation measures are not effective in most areas.

THE IMPACT OF THE GANGA ACTION PLAN (GAP): As with previous species, pollution abatement schemes will definitely have a positive impact on the river water, and ultimately, on the habitat of the animal.

North Indian Roofed Turtle

The North Indian Roofed Turtle (*Kachuga tecta*) is categorized as vulnerable by the Zoological Survey of India. As before, the species is found in the Ganga and Brahmaputra river systems in India. This species is seldom exploited for food, generally being too small. Some numbers are believed to be caught for the pet trade in the west originating from Bangladesh, and perhaps, also Pakistan (Das, 1991). Eggs are exploited for food and the habitat is being destroyed due to the construction of dams/barrages on the rivers, removal of sand from banks, and the loss of vegetation cover all along the flood plains of the river. This has reduced nesting space and food availability.

It is included under Schedule I of the Indian Wildlife (Protection) Act, 1972 and in Appendix I of CITES (Choudhury and Bhupathy, 1993). This common species is given the strictest legal protection (Das, 1991).

THE IMPACT OF THE GANGA ACTION PLAN (GAP): The GAP will improve the water quality of the Ganga and its tributaries, which in turn, will have a positive impact on the species.

Three-striped Roofed Turtle

The Three-striped Roofed Turtle (*Kachuga dhongoka*) is considered extremely vulnerable. It is found throughout the Eastern Ganga River System including all its major tributaries, as far west as Allahabad, and in the north up to Nepal. The animal is heavily exploited for its flesh. Eggs are also exploited for food. Habitat destruction due to Water Development Projects and aquatic pollution pose threats (Das, 1991). However, no specific data on the effects of such habitat modifications are available. It is categorized under Schedule IV of the Indian Wildlife (Protection) Act, 1972, and is generally protected by legislation, but the conservation measures are not effective.

THE IMPACT OF THE GANGA ACTION PLAN (GAP): The GAP will improve the water quality of the Ganga and its tributaries, which in turn, will have a positive impact on the species.

Red-crowned Roofed Turtle or Bengal Roofed Turtle

The Red-crowned Roofed Turtle or Bengal Roofed Turtle (*Kachuga kachuga*) is categorized as extremely vulnerable by the Zoological Survey of India and endangered by the IUCN, and is distributed in the main stem of the Ganga and its tributaries in Uttar Pradesh, Bihar, and West Bengal. It is occasionally exploited for its flesh (Das, 1991). Further threats come from habitat destruction and abrupt change in ecology due to the removal of sand or earth from the banks of the river for commercial purposes, leading to loss of aquatic as well as riparian vegetation, creating scarcity of food and destroying nesting sites. The species is included under Schedule I of the Indian Wildlife (Protection) Act, 1972, but conservation measures are not adequate. A small-scale project has been undertaken since the late 1980s for the artificial incubation of eggs collected from wild nests at a few places along the National Chambal Sanctuary.

THE IMPACT OF THE GANGA ACTION PLAN (GAP): Except for improvement in water quality due to pollution abatement schemes, no direct impact is visualized.

River Terrapin

The River Terrapin (*Batagur baska*) is categorized by the IUCN as endangered. Prehistoric remains indicate that it was widely distributed in the Ganga river system up to the twelfth century AD. Currently the River Terrapin is restricted to the estuaries of the Ganga and Brahmaputra, known as the Sundarbans. It is a monotypic genus (Das, 1991). Threats to the species are posed by over-exploitation of the eggs and flesh for food. In mid-nineteenth century Bengal, its fat was widely used for the manufacture of soap. Habitat destruction due to mining, removal of river-side sand, clearing of riverside vegetation, and exposing the bank to erosion, construction of dams and barrages, urbanization along the banks, and an increase in steamer traffic have led to a decrease in the terrapin population. It is protected under Schedule I of the Indian Wildlife (Protection) Act 1972 and Appendix I of CITES, indicating that all forms of exploitation are banned. During the last 35 years, several hatcheries have been established in Malaysia to incubate Batagur eggs. About 10,000 hatchlings have been released in the Perak River since 1969 (Tikader and Sharma, 1985).

THE IMPACT OF THE GANGA ACTION PLAN (GAP): No direct impact of the GAP on the River Terrapin has been observed. However, a reduction in pollution will, in turn, help in improving the water quality in the estuarine zone of the river.

SPECIES (OTHER THAN THOSE IN THE FISHERY STATISTICS) IN THE GANGA RIVER OR BANK SIDE, WHICH HAVE SIGNIFICANT COMMERCIAL USE VALUE

There are many species in the Ganga, not solely fish, which have significant commercial value. Otters, crocodiles, and turtles are killed mainly for commercial purposes. Most of these species—especially otters and crocodiles—are in high demand in the international market, and thus, have worldwide commercial value. Otters are killed for their skin, the fur of which is in very high demand all over the world. Similarly, the hide of crocodiles is also in great demand in the international market. Soft-shelled turtles contain rich protein in their flesh. These animals are also in high demand, both in the national and international market. Some body parts of most of these animals are used for medicinal purposes. Some of the hard-shelled turtles have commercial value as they are used as pet animals, especially in the West. These species have been discussed in the previous section.

Besides aquatic megafauna, there are many wild plants and animals along the banks of the river which have economic value and commercial use.

A few decades ago, on the bank side of the Ganga, a diversified natural vegetation was very common. Some of the vegetation like *Saccharum arundaceum* and *Saccharum spontaneum* are still harvested for economic gain. The local riparian people of low strata make mats of the leaves of *Saccharum arundaceum* and also use it for thatching their houses and making baskets. Though no data on the amount harvested is known, it provides subsistence to thousands of people in the upper-middle reaches of the Ganga and its tributaries in Uttar Pradesh. Similarly, Saccharum Spontaneum is commercially harvested for thatching, fuel, and fodder by riparian people, especially in eastern Uttar Pradesh and Bihar. No specific data is available on the amount harvested, commercial value, population status, and sustainability, but the area covered with this vegetation is fast declining every year due to encroachment by people for farming on riverbank land. Moreover, such vegetation provided shelter to varieties of wild animals like leopards, wild pigs, blue bulls, etc. The leopard (*Panthera pardus*) is almost extinct on the river bank. In the recent past, one leopard was caught in 1993 on the Ganga flood plain vegetation near Bakhtiarpur in Bihar. Wild pigs (*Sus scrofa*) have also become extinct from the bank of the Ganga.

The blue bull (*Boselaphus tragocamelus*) can still be seen in the Saccharum cover and, also in the open on the bank of the river. This animal is often killed for its meat, but no data is available for the quantity harvested, its commercial value, population status, or sustainability. A few decades ago, the population of this animal, as reported by locals, was plentiful, but due to over-exploitation and loss of vegetation cover from the bank side, the population has declined to a very low level, and now they are found only in a few areas and with very low density. This animal is also almost on the verge of extinction from the river bank.

In general, the impact of the pollution of the Ganga and GAP activities on terrestrial plant and animal species of commercial use value using the riverine habitat will be relatively small compared to direct terrestrial impacts such as land-use change and over-harvesting.

OVERALL ASSESSMENT OF 'ECOSYSTEM HEALTH' NOT CAPTURED ELSEWHERE

The River Ganga has been a symbol of purity since time immemorial and has served as the cradle of Indian civilization: the life-line of the people of India. Ancient Indian treatises on medicine and dietetics from the *Charaka Samhita* (first century AD) onwards, are unanimous in their description of the Ganges as wholesome, clear, sweet, tasty, and digestive.

The centuries-old exceptional qualities of Ganga water were, however, threatened, with the increasing rate of the discharge of untreated toxic industrial effluents, as well as urban domestic sewage into the rivet all along its course. The contaminated river water became unfit, both for drinking and bathing at all major cities. Not only has the pristine purity of the Ganga water deteriorated, but the entire ecology of the river has been degraded. Fishing fields of commercial importance, viz. Indian major carps, hilsa etc., collapsed; the monoculture of pollution tolerant-species like tubificids became very common, especially near city outfalls; pollution intolerant (sensitive) species disappeared; community respiration increased; and primary productivity of the river was adversely affected, resulting in the decline of not only the fish population, but of other vertebrates as well. Indiscriminate use of organochlorine pesticides in agriculture and the health sector posed new threats and resulted in a high accumulation of these hazardous chemicals in the tissues of fish as well as of other higher vertebrates due to biomagnification. Finally, these chemicals are gaining entry into the human body through several different pathways. Rampant killing of Ganga turtles, especially the Soft-shelled Turtles (*Aspideretes gangeticus*, *Lissemys punctata*, etc.) by fishermen reduced the scavenging capacity of the river system, ultimately adversely affecting the self-purifying capacity of the river. *Aspideretes gangeticus* is a very good scavenger and feeds mainly on dead bodies and carcasses. Direct and accidental trapping and habitat destruction due to various developmental activities, as well as pollution, pushed the only cetacean in the Ganga, the Ganga River Dolphin, to the verge of extinction. The river has thus, in the last few decades, reached its worst ever state of deterioration and degradation.

Undoubtedly, the various schemes under the Ganga Action Plan have improved the physico-chemical quality of the river and have had positive effects on the biota of the river too. Some indications of the return of some threatened species are evinced within the short span of only one decade. The Ganga Action Plan does not have an ecological restoration component, nevertheless, the improvement in water quality has probably resulted in the return of biota. It should be noted that in most cases the improvements in Ganga pollution brought about by the GAP were in areas where the river was so polluted that rare and endangered species were already absent, and these areas would never again be their prime habitat due to population pressure and disturbance. However, it is important that rare species should be able to move through these regions in order to allow recolonization of better areas and gene flow between isolated populations.

The 'ecosystem health' can be assessed overall on the basis of some specific biological parameters, viz. bacteriological load, primary productivity, the number of species and density of phytoplankton and zooplankton, as well as benthic macroinvertebrates, fish catch, etc.

As far as time series data on important general indicator species or ecosystem health is concerned, this is very fragmented, meagre, and piecemeal. Research has been conducted at various centres along the Ganga according to the expertise available and the immediate objectives of the investigations. In this study, an effort has been made to collect and collate the data available from various sources and literature.

PRIMARY PRODUCTIVITY

Compared with phytoplankton, zooplankton, and benthic macroinvertebrates, data on primary productivity is very meagre. Jhingran (1991) observed a significant improvement in the water quality and primary productivity rate at Kanpur due to the positive impact of diversion and treatment of the effluents, prior to their release in the river, since 1987. He found a consequent increase in the fish production potential of the stretch, from 8 to 144 kilograms per hectare per year to 111 to 184 kilograms per hectare per year. The detailed data is shown in Table 11.1.

Rao (1996) recorded primary productivity at eight different centres between Rishikesh and Kanpur during 1993–4. The results are shown in Table 11.2.

PHYTOPLANKTON

The data for Phytoplankton density were taken from a variety of sources and are summarized in Table 11.3. Pahwa and Mehrotra (1966) undertook an investigation in 1960 on plankton (both phyto- and zooplankton) at Kanpur (A), Allahabad (B), Varanasi (C), Ballia (D), Patna (E), Bhagalpur (F), and Rajmahal (G), a stretch of 1090 kilometres. The maximum densities of phytoplankton at centres A to G were 77,870; 42,929; 24,092; 7601; 1353; 1114; and 739 units per litre respectively during the summer months (May and June).

TABLE 11.1

Change in Primary Productivity in the River Ganga at Kanpur

Zone	Before diversion				After diversion			
	1	2	3	4	1	2	3	4
Fish production potential (tons)	4152	2968	3913	222	4352	3212	5309	5256
Photosynthetic efficiency (%)	0.355	0.254	0.330	0.019	0.372	0.272	0.454	0.45
Energy fixed by producers (cal/m²/day)	144	103	136	8	151	111	184	182

1 Bhagwatghat above confluence 3 Jajmau above confluence
2 Bhagwatghat below confluence 4 Jajmau below confluence

TABLE 11.2

Primary Productivity of the River Ganga at Different Sites Between Rishikesh and Kanpur, 1993–4

	Rishikesh	Hardwar	Bijnor	Brijghat (Garhmukteswar)	Narora	Kachlaghat (Badaun)	Ghatiaghat (Farrukhabad)	Kanpur
GPP (mg C/m²)	104.6	107.7	105.7	97.3	96.9	105.8	105.1	103.8
NPP (mg C/m²)	30.85	31.7	31.0	24.6	20.5	18.7	17.8	16.1
Respiration (mg C/m²)	21.6	22	22	24.5	26.6	30.6	31.8	37

TABLE 11.3

Phytoplankton Density (unit/l) in the River Ganga at Different Sites

	Kanpur	Allahabad	Varanasi	Ballia	Patna	Bhagalpur	Rajmahal	Between Bhagalpur and Lalgola
1960	77,870	42,929	24,092	7601	1353	1114	739	na
1970	na	na	na	na	na	2107–4666	na	na
1971	na	na	na	na	na	na	na	na
1972	na	na	na	na	na	1639	na	153–4857
1973	na	na	na	na	na	na	na	126–4004
1974	na	na	na	na	na	na	na	160–5188
1975	na	na	na	na	na	na	na	105–5826
1976	na	na	na	na	na	na	na	204–42,276
1977	na	na	na	na	na	412–22,824	na	na
1978	na	na	na	na	na	162–3084	na	na
1981	na	na	na	na	na	32–740	na	na
1982	na	na	na	na	na	48–1294	na	na
1984	na	na	na	na	na	76,000	na	na
1978–9	na	na	na	na	na	1344	2718	na
1980s	59,628	na	8300	8300	296	1721	1730	na
1990s	1189	949	1569	1450	950	2891	1705	na

na = not available

The dominant groups of phytoplankton were found to be *Bacillario-phyceae, Chlorophyceae,* and *Myxophyceae.* Reinhard (1931), Roy (1955), and Shetty (1961) confirmed this at the Kanpur, Allahabad, and Varanasi centres, but found that at Ballia, Patna, and Rajmahal the annual crop of phytoplankton was clearly dominated by the green algae. At Bhagalpur the diatoms and green algae were present in almost equal numbers. The blue-green algae formed a sizeable portion of the phytoplankton only at Kanpur, Allahabad, and Varanasi. At other centres they were least abundant. Diatoms, represented by *Synedra* spp and *Fragillaria* spp ,dominated river stretches at Kanpur and Allahabad; Chlorophyceal, represented by *Mougeotia, Spirogyra,* and *Pediastrum,* were dominant at Bhagalpur. CIFRI (1985) scientists recorded phytoplankton density at Bhagalpur. During 1970, 1971, 1977, 1978, 1981, 1982, and 1984, phytoplankton density was 2107 to 4666, 1639, 412 to 22,824, 162 to 3084, 32 to 740, 48 to 1294 and 76,000 units per litre respectively. Between Bhagalpur and Lalgola the phytoplankton density during 1972, 1973, 1974, 1975, and 1976 was 153 to 5857, 126 to 4004, 160 to 5188, 105 to 8826, and 204 to 42276 unit per litre respectively.

In the late 1970s (1978–9), Bhagalpur University carried out an investigation in the Barauni–Farakka segment of the Ganga and recorded 1344 and 2718 units per litre of phytoplankton at Bhagalpur and Rajmahal respectively (Bilgrami and Datta Munshi, 1979). In the 1980s, 14 different universities carried out investigations under the GAP at Kanpur, Varanasi, and Ballia, and densities of 59,628; 830,000; 830,000 units per litre respectively of phytoplankton were recorded. During the same period, the density of phytoplankton at Patna, Bhagalpur, and Rajmahal was 296, 1721, and 1730 units per litre respectively (Krishnamurti *et al.,* 1991). In the mid-1990s, the phytoplankton density at Kanpur, Allahabad, Varanasi, and Ballia was 1189, 949, 1569, and 1450 units per litre (Khan, 1996). Bilgrami (1992) reported phytoplankton density of 2891 and 1705 units per litre at Bhagalpur and Rajmahal respectively during January 1990 to December 1991. During the 1990s, the phytoplankton density at Patna was 950 units per litre.

Rao (1996) recorded phytoplankton abundance at different centres between Rishikesh and Kanpur during 1993–4. The density of phytoplankton at Rishikesh, Hardwar, Bijnor, Brijghat (Garhmukteshwar), Narora, Kachlaghat (Badaun), Ghatiaghat (Farrukhabad), and Kanpur was 707, 819, 732, 766, 844, 885, 1097, and 1233 units per litre respectively during 1993–4.

ZOOPLANKTON

The data for zooplankton density were taken from a variety of sources and are summarized in Table 11.4. Pahwa and Mehrotra (1966) recorded the density of zooplankton as 2821, 2684, 1033, 105, 26, 183, and 369 units per litre at Kanpur, Allahabad, Varanasi, Ballia, Patna, Bhagalpur, and Rajmahal respectively during their investigation in 1960. Rotifers and copepods dominated the zooplankton populations throughout the year. The contribution of zooplankton to the total plankton crop was only 4.8 per cent, 5.3 per cent, 11.3 per cent, 5.1 per cent, 8.3 per cent, 8.7 per cent, and 14.0 per cent at Kanpur, Allahabad, Varanasi, Ballia, Patna, Bhagalpur, and Rajmahal respectively. *Brachionus* and *Keratella* spp were the most dominant rotifer genera and together contributed 77.6 per cent, 57.5 per cent, 49.3 per cent, 71.9 per cent, 75.0 per cent, 63.2 per cent, and 80.4 per cent to the total rotifer crop at the centres from Kanpur to Rajmahal respectively.

Pahwa and Mehrotra (1966) concluded that the River Ganga at Allahabad was more productive above its confluence with the Yamuna. This conclusion has been supported by CIFRI scientists.

CIFRI scientists recorded zooplankton density at Bhagalpur. In the years 1970, 1971, 1977, 1978, 1981, 1982, and 1984 the density was 423, 199, 128, 23 to 276, 120 to 576, 6 to 336, and 10 to 400 organisms per litre respectively. Between Bhagalpur and Lalgola, zooplankton density was 5 to 217, 4 to 169, 2 to 57, 3 to 36, and 4 to 40 during 1972, 1973, 1974, 1975, and 1976.

The data on the zooplankton community is meagre. During 1978–9, the zooplankton density at Bhagalpur and Rajmahal was only 17 and 18 units per litre respectively (Bilgrami and Datta Munshi, 1979). During the 1980s the zooplankton data was collected at Buxar, Patna, Bhagalpur, and Rajmahal and the density was recorded to be 125, 199, 53, and 55 units per litre respectively (Sinha, and Prasad, (eds), 1988; and Bilgrami and Datta Munshi, 1985). In the 1990s (1993–4) the zooplankton density at Kanpur, Allahabad, and Varanasi was 329, 376, and 369 units per litre respectively (Khan 1996). Regular data on zooplankton was recorded between 1991 and 1994 at Buxar, Patna, Bhagalpur, and Rajmahal (Sinha, 1996).

The composition of plankton has not changed much since 1960, but abundance has increased significantly in the upper and middle reaches of the Ganga (Natarajan, 1989).

During 1993–4, zooplankton abundance at Rishikesh, Hardwar,

TABLE 11.4

Zooplankton Density in the River Ganga at Different Sites

(Organisms/l)

	Kanpur	Allahabad	Varanasi	Buxar	Ballia	Patna	Bhagalpur	Rajmahal	Between Bhagalpur & Lalgola
1960	2821	2684	1033	na	105	26	183	369	na
1970	na	na	na	na	na	na	423	na	na
1971	na	na	na	na	na	na	99	na	na
1972	na	na	na	na	na	na	na	na	5–217
1973	na	na	na	na	na	na	na	na	4–169
1974	na	na	na	na	na	na	na	na	2–57
1975	na	na	na	na	na	na	na	na	3–36
1976	na	na	na	na	na	na	na	na	4–40
1977	na	na	na	na	na	na	128	na	na
1978	na	na	na	na	na	na	23–276	na	na
1981	na	na	na	na	na	na	12–576	na	na
1982	na	na	na	na	na	na	6–336	na	na
1984	na	na	na	na	na	na	10–400	na	na
1978–9	na	na	na	na	na	na	17	18	na
1980s	na	na	na	125	na	199	53	55	na
1990s	329	376	369	na	na	86	112	118	na
1991	na	na	na	90	na	146	66	104	na
1992	na	na	na	108	na	116	90	102	na
1993	na	na	na	96	na	82	148	78	na
1994	na	na	na	86	na	na	na	na	na

na = not available

Bijnor, Brijghat (Garhmukteshwar), Kachlaghat (Badaun), Ghatiaghat (Farrukhabad), and Kanpur was recorded to be 65, 72, 83, 87, 106, 111, and 148 units per litre respectively (Rao, 1996).

During the investigation carried out between 1982 and 1995, an increase in the number of species as well as the density of zooplankton was recorded at Buxar and Patna. During 1985–7, the total number of zooplankton species was 27, which rose to 56 in 1991–5. The zooplankton density (units per litre) increased from 83 in 1982–7 to 109 in 1995 and from 83 in 1985–7 to 92 in 1995 at Patna and Buxar respectively.

In the lower Ganga, both plankton composition and abundance showed remarkable changes after the construction of the Farakka barrage. Plankton count ranged from 37 to 95 per litre during the pre-barrage years (before 1975), but increased to 145 to 304 per litre in the post-barrage year, 1980 (Natarajan, 1989). During 1960, estuarine forms of the phytoplankton such as *Coscinodiscus cranii, Asterionella sapponica, Chaetoceros* spp, and zooplankton *Pseudodiaptomus* spp *Microsetella* dominated the tidal stretch from Calcutta to Diamond Harbour. After barrage construction, freshwater forms became abundant as far downstream as Diamond Harbour. In this area, *Chlorophycaea* has increased by an order of magnitude between 1960 and 1980.

BENTHIC MACROINVERTEBRATES

The data for benthic macroinvertebrates density were taken from a variety of sources and are summarized in Table 11.5. Pahwa (1979) carried out an investigation on benthic macroinvertebrates for a stretch of 1090 kilometres of the Ganga between Kanpur and Rajmahal in 1960. He recorded an average density of benthic macroinvertebrates of 3476, 1121, 880, 71, 1593, 218, and 508 individuals per square metre at Kanpur, Allahabad, Varanasi, Ballia, Patna, Bhagalpur, and Rajmahal respectively.

Insects belonging to the order Diptera, Ephemeroptera, Odonata, Hemiptera, Trichoptera, and Coleoptera were predominant at Kanpur, Allahabad, and Varanasi, contributing 94.19 per cent of the total benthic fauna. At the time of the study, Ephemeroptera and Trichoptera, which are sensitive to organic pollution, were present even at Kanpur, the most polluted stretch of today's Ganga, and constituted 2.36 per cent and 11.52 per cent respectively of the total insect fauna. On the other hand, Annelida (Oligochaeta), the most pollution tolerant group of benthic macroinvertebrates, were absent and only leeches, forming 1.59 per cent of total benthos, were present at Kanpur. It shows that the Ganga at Kanpur in 1960 was relatively pollution free. The average abundance of benthos was as high as 3476 per square metre (insects 3405 per square metre) at Kanpur but decreased to 218 per square metre at Bhagalpur (molluscs 124 per square metre).

TABLE 11.5

Benthic Macroinvertebrates Density in the Ganga River at Different Sites

(Individual per square metres)

	Kanpur	Allaha-bad	Varanasi	Ballia/Buxar	Patna	Bhagal-pur	Raj-mahal	Farakka
1960	3476	1121	880	71	1593	218	508	na
1970s	na	na	na	na	na	na	na	na
1980s	na	na	na	728	277	na	na	na
1990	na	na	na	na	na	na	na	na
1991	na	na	na	953	1735	1735	10023	1669
1992	na	na	na	4414	na	3418	3447	4918
1993	1311	1008	1175	3616	3645	2534	6096	4127
1994	1146	521	993	3606	4201	2619	10866	12316
1995	na	na	na	4514	4985	2812	na	12825

na = not available

The time series data on benthic macroinvertebrates is meagre compared with phytoplankton and zooplankton. During the 1980s, 728 and 277 individual square metres benthic macroinvertebrates were recorded at Buxar and Patna respectively (Sinha and Prasad (eds), 1988). During 1982–95, an increase in the number of species and density of benthic macroinvertebrates was recorded. The number of species increased from 22 in 1985–7 (Sinha and Prasad (eds), 1988) to 30 in 1991-5 (Sinha, 1996). At Buxar, the production of benthic macroinvertebrates increased from 728 per square metre in 1985–7 to 4514 per square metre in 1995, whereas at Patna it increased from 277 per square metre to 4985 per square metre during the same period. At other places like Munger, Sultanganj, and Farakka, the increase was from 2217 in 1992 to 5538 in 1995, from 519 in 1991 to 2812 in 1995, and from 1669 in 1991 to 12,825 in 1995 respectively. The monoculture-like situation of tolerant species like tubificids (10^6 per square metre) during the early 1980s at city outfalls disappeared in the mid-1990s, indicating an improvement in the ecosystem health. Regular data on benthic macro-invertebrates in the Buxar–Farakka segment of the Ganga for a stretch of about 600 kilometres have been collected between 1991 and 1994 (Sinha, 1996).

Falls in the density of benthic macroinvertebrates at Kanpur, Allahabad, and Varanasi were recorded between 1993 and 1994 (Khan, 1996).

Recent studies in the lower Ganga (CIFRI, 1985) show a sharp fall in the density of benthic macroinvertebrates which may be attributed to paper pulp and sewage sludge. Maximum numbers (94–164 per square metre) were noted at Nabadwip, while minimum numbers were 3 to 15 per square metre.

MIGRATORY FISH (HILSA)

Prior to 1948, virtually no information on riverine fish catch statistics was collected systematically. However, some sparse information is available. Dr S.L. Hora, the then Director of the Zoological Survey of India, in his note to the Planning Commission of India, which was later published in the Journal of the Asiatic Society in 1952, gave some catch statistics of fish in different states. In Uttar Pradesh, Bihar, and West Bengal the total estimated catch was 15,700; 1,009,500; and 1,197,800 mounds (one mound = approximately 37 kilograms) respectively (Hora, 1952). It is not known if this catch is only from the Ganga or from all freshwater bodies.

Detailed information on the fishing potential and catch statistics was collected by the CIFRI from 1956 onwards. The Riverine Division of the Institute located at Allahabad collected regular data for the river Ganga from Bulandshahar in Uttar Pradesh to Lalgola in West Bengal, a stretch of about 1580 kilometres.

Khan (1996) reported that at Allahabad, production of *hilsa ilisha* in 1961–8 was 10.38 kilograms per hectare which declined to 0.13 kilograms per hectare in 1989–93. Similarly at Buxar, the average annual catch of hilsa decreased from 33.48 tonnes in 1963–71 to only 0.68 tonnes in 1973–8. At Patna, the production of hilsa was 9.39 kilograms per hectare in 1961–6, and decreased to 0.07 kilograms per hectare during 1989–93. At Bhagalpur the hilsa catch during 1961–70 was 4.27 tonnes per annum and this declined to 0.79 tonnes in 1985–8. Natrajan (1989) recorded that in the Allahabad stretch of the Yamuna, the catch rate for hilsa decreased from 0.91 to 0.08 kilograms per hectare per year over the same period. Natrajan (1989) emphasized that the structure of fishery has changed drastically, with high-value major carps and hilsa giving way to low value species. The anadromous hilsa, formerly a significant catch in the Ganga system, has been all but eliminated from the middle stretches.

Some significant positive changes in the hilsa catch have been seen at certain locations in the last few years. At Patna, the hilsa catch in 1988, 1989, 1990, and 1991 was 0.11, 0.04, 0.01, and 0.03 metric tonnes respectively, whereas the catch during 1992, 1993, 1994, and 1995 became 0.10, 0.36, 0.40, and 0.65 tonnes respectively. This represents a 600 per cent increase between 1988 to 1995.

If one analyses the overall ecosystem health of the Ganga, it is apparent that many factors are operating simultaneously in the system.

The annual run-off in the Ganga basin after allowance for evapotranspiration, soil seepage, and ground water recharge, is about 469 billion cubic metres. Of this, 85 billion cubic metres of water is diverted by canal projects and by hydroelectric and storage reservoirs for irrigation, power, and flood control. Canal projects account for a little over 60 per cent of the impounded water. The diversion of water has caused large-scale changes in the channel bed and hydrography of the river in terms of flow, flow-rate, flood rhythm, and regime. Hydraulic structures have changed river morphometry, increased bank erosion and created barriers for migratory fishes. The construction of flood-control dykes and levees in flood-prone low-lying areas has deprived the major carps of their extensive breeding habitats.

About 62 per cent of the Ganga basin is under cultivation. At present, agriculture uses annually about 1.5×10^6 tonnes of chemical fertilizers, mostly N-based, and 2600 tonnes of pesticides, a good proportion of which are non-degradable organochlorines. About 27 billion cubic metres of waste water carrying toxic pesticide residues and chemical nutrients from fertilizers drains from irrigated fields into the river system (Das Gupta, 1984).

Reduction in forest cover has led to great increase in both the dissolved and particulate solids in the Ganga river system (Rao, 1976). The turbidity level has increased and the sediment load has reduced the effective photosynthetic zone and diminished productivity.

While agriculture, power production, industrial development, flood control, domestic water supply, and navigation receive priority consideration, the basic water requirements necessary to protect breeding habitat and to provide summer shelter for fish brood stocks have received low priority.

Changes in hydrography due to canal projects and flood control measures, and changes in river morphometry have reduced considerably the extent of breeding habitat and have depressed recruitment of major carps.

The long range migrant, *hilsa ilisha*, forms the basis of an important fishery in the Ganga River System with an annual production of about 1500 tonnes from the delta and up to 100 tonnes from the middle reaches. The fish is vulnerable to any hydraulic structure that impedes its migrations, particularly in the deltaic stretches of the river. The natural migratory range of the fish is 1500 kilometres from the Hooghli estuary to Allahabad on the Ganga. The 1975 construction of a barrage at Farakka, at the head of the Bhagirathi and Padma tributaries of the Ganga, some 470 kilometres from the river mouth, has adversely affected the hilsa fishery in the middle reaches and has nearly eliminated the riverine fishery upstream of Farakka on the mainstem of the Ganga.

Because of fishing, there is large-scale destruction of brood stocks of the Hooghly hilsa in the feeder canal and near the head regulator of the feeder canal at Farakka.

CONCLUSIONS

This chapter has examined the impacts of the GAP on the biodiversity of the Ganga ecosystem. It has looked at three areas:

- Species of international conservation significance
- Species that have commercial significance
- Overall assessment of 'ecosystem' health.

There are a number of species of international significance in the Ganga river and along its banks. They include:

- Mammals (Ganga River Dolphin, Irrawady Dolphin, Common Otter, Smooth-coated Indian Otter, and Asian Small-clawed Otter)
- Birds (Large Whistling Teal, Osprey, 'Pallas' Fish Eagle, and Indian Skimmer)
- Reptiles (Gharial, Marsh Crocodile, Saltwater Crocodile, Indian Soft-shelled Turtle, Peacock-marked Soft-shelled Turtle, North Indian Flap-shelled Turtle, Narrow-headed Soft-shelled Turtle, Black-spotted Pond Turtle, Brahminy Turtle, North Indian Roofed Turtle, Three-striped Roofed Turtle, Bengal Roofed Turtle, and River Terrapin).

The benefits of the GAP on these species arise in four ways. First, there are some species for which there have been captive breeding programmes. This applies to the Ganga Dolphin, Ganga Soft-shelled Turtle and the North Indian Flap-shelled Turtle, the Gharial, the Marsh Crocodile, and the Saltwater Crocodile. The survey does not comment in detail on the success of those programmes, but gives the impression that they have helped the conservation effort. Second, the GAP has raised awareness and encouraged conservation efforts through information dissemination etc. This has been important for the Ganga Dolphin, although it is difficult to quantify the impact. Third, GAP has facilitated the collection of information on species and their habitat, and this will contribute in an important way to their conservation. Beneficiaries of these efforts have been the Ganga Dolphin, the Gharial, and the other species for which conservation programmes have been initiated. Finally, the general improvement of the water in the Ganga has helped most of the above species, except those whose habitat is brackish or saltwater. The extent of the impact of the reduced organic and toxic pollution loads is, however, not available.

The commercial species looked at in this chapter exclude commercial fisheries, which are dealt with in Chapter 9. Otters, turtles, and crocodiles are killed for commercial purposes. In addition, there is a diversified natural vegetation base that is harvested for personal and commercial gain, especially local reeds, used for basket-making, thatching, etc. The area under this vegetation is fast disappearing and there is no indication that the GAP has reversed this trend. Another species that is disappearing is the blue bull, an animal killed for its meat. In general, the GAP has had little impact on these species, which are endangered because of land-use change and over-harvesting.

The overall ecosystem health is assessed in terms of specific biological parameters, namely, the primary productivity, phytoplankton, zooplankton, benthic macroinvertebrates and migratory fish.

Primary productivity is measured in terms of the fish production potential, the photosynthetic efficiency of the river, and the energy fixed by producers. Although data are scarce, there is some evidence, from studies at eight sites between Rishikesh and Kanpur, that primary productivity has improved since the diversion and treatment of effluents began on that section.

Phytoplankton data have been compared between the 1980s and the 1990s and show a mixed picture. Improvements have been recorded for Bhagalpur and Patna, but declines have been noted in Kanpur, Varanasi, and Ballia. No explanation is available for these trends.

Zooplankton data are sparse. However, increases between the 1980s and 1990s have been observed in Buxar, Patna and in the lower Ganga, below the Farakka barrage. The significance of the these is not altogether clear, except that it is a positive indicator of the ecosystem health.

Benthic macroinvertebrate data are even more meagre than those for zooplankton, but this study has assembled what was available. They show an increase at Buxar, Patna, Bhagalpur, Rajmahal, and Farakka. Recent studies have shown, however, a sharp fall in macroinvertebrate density in the lower Ganga, which is attributed to pulp and paper sewage and sludge.

The migratory fish (hilsa) has had considerable information collected on it since the 1950s. Substantial declines in catch were recorded in periods from the early 1970s to the early 1980s (Allahabad, Buxar, and Bhagalpur), and declines were recorded in the late 1980s and early 1990s (Patna and Lalgola). Subsequently there has been some increase in the catch. At other landing stations the trends are not clear.

All these factors point to a situation that is slightly improved with respect to the early 1980s. Nevertheless, long term trends on ecosystem health are influenced by many factors, including pesticide loading, riverine forest cover, and hydraulic structures that change river morphometry, increase bank erosion, and create barriers for migratory fish. These trends continue to operate, and increasingly, to damage the ecosystem.

12

Economic and Financial Evaluation of the Ganga Action Plan

INTRODUCTION

In this chapter we bring together the monetary estimates of costs and benefits from the earlier chapters, to analyse the net benefits of the GAP. The net benefits are calculated for the following cases:

- Benefits as a result of the water quality improvements actually achieved in the period 1985–95 (referred to as GAP I);
- Benefits as a result of the total benefits expected, once all stages of the GAP are completed (referred to as GAP I and GAP II);
- Benefits as a result of the incremental gains in net benefits, in going from where we are now with the GAP, to the completion of the goals of the GAP, that is in implementing GAP II.

The net benefits are reported in terms of net present values, using a real discount rate of 10 per cent, as well as in terms of the internal rate of return.[1]

The analysis is carried out in terms of the actual financial costs, as well as in terms of the economic costs. The financial costs are all reported in 1995 prices and in Indian Rupees. The economic costs are also reported in 1995 prices, but in contrast with the financial costs, shadow prices have been applied, to allow for distortions in the market prices and income distribution weights are used to capture the income distribution benefits of the GAP.

The following section reports and discusses the cost and benefit estimates, while the subsequent sections present the results of the finan-

---

[1] A real rate of discount of 10 per cent is taken as one that is used as the cut-off rate for public sector projects appraised in India. Any further premium on capital is reflected in the shadow price on capital expenditures.

cial analysis, the results after applying shadow prices, an analysis of income distribution benefits of the GAP, the financial sustainability of the GAP, and finally, some concluding comments.

COSTS AND BENEFITS

COST DATA

The cost data were discussed in detail earlier. GAP projects are divided into those associated with GAP I—the capital expenditures for which have effectively been completed—and GAP II, which was started in 1993 and will continue for a few more years. The investment expenditure at constant prices for GAP I was Rs 7 billion and for GAP II will be Rs 4.2 billion, of which Rs 6.11 billion have been spent by the end of 1996. Annual details of the investment expenditures are given in Table 4.1. The investment cost data are broken down into domestic materials and equipment, skilled labour, and unskilled labour. This breakdown is necessary when we estimate the economic costs and different shadow prices apply to the different categories of costs. The same categories apply to the operating costs.

SHADOW PRICES

When the market prices are considered not to reflect social values, adjustments are made through the use of 'shadow prices'. These are coefficients by which the market prices are multiplied to obtain the social prices. There are several adjustments that can be made (for details see Dasgupta, Marglin and Sen, 1972; Little and Mirrlees, 1974). Some correct for market failures and some are applied to correct for the different social values associated with income by different sections of society. In the adjustments made in this report, the corrections to market prices reflect primarily two factors: (a) the scarcity of capital that is greater than its market value would suggest; and (b) the lack of employment opportunities for unskilled labour, so that the opportunity cost of such labour is lower than of the wages paid on the profit. Based on work by Murty *et al.*, (1992), the shadow prices used have been the following:

        capital materials and equipment: 1.4
        unskilled labour: 0.5

The effects of using these shadow prices are shown in Table 12.1. The table gives the present value of capital and operating costs of the project, both in financial and economic terms. The economic costs are 6 to 16 per cent higher for the investment category and about 24 per cent higher for the operating costs. The effects of the undervaluation of capital and materials costs dominate the overvaluation of unskilled labour. Nevertheless, as will be seen below, the effects of using economic costs on the conclusions of the analysis is not large.

TABLE 12.1

Financial and Economic Costs in Present Value Terms

Discount rate used is 10 per cent   (Rs mn)

Cost category	Financial cost	Economic cost
GAP I—Investment	4245	4918
GAP I—Operating	324	422
GAP II—Investment	2716	2893
GAP II—Operating	192	242

BENEFITS

The categories of benefits considered for monetary estimation were: non-user benefits, user benefits, health, toxics, fisheries, and agriculture. Of these, some estimates were successfully obtained for the non-user and user categories, health, and agriculture. Satisfactory estimates could not be obtained for fisheries benefits and toxic impacts (which are partly benefits and partly costs; see chapter 8). The details on the four sets of monetized benefits are given in the respective chapters dealing with those impacts. The key numbers used from those chapters in the final analysis are summarized here.

Non-user benefits

The key table for non-user benefits is Table 5.10. As noted at the beginning of this chapter, there are three kinds of improvements analysed: the value of the present improvements in water quality, the value of all improvements expected once both phases of GAP have been realized, and the value of the additional improvement in going from present quality to the full quality goals. In comparing the present quality with the quality before the GAP there is the choice of taking the 1985 quality levels or taking the levels that would have prevailed in 1995 had GAP not been implemented. As noted in Appendix A and Chapter 5, the water quality in 1995 would have been worse than the 1985 levels in the absence of the GAP.

Table 12.2 sums up the main non-user benefits. In comparing the 1985 quality with the 1995 or full GAP quality there is a range because the model used has a key variation in parameters that gives the lower and upper bounds of that range. If we compare the estimate of annual benefits of 'Full GAP' quality with the quality that would exist without the GAP, the estimate is higher than the corresponding estimate when the 1985 quality is taken. The difference is between 5 and 25 per cent, depending on which comparison is taken.

TABLE 12.2
Summary of Non-user Benefits Utilized in Analysis

(Rs million 1995)

Changes in water quality	Estimated annual benefits in 1995
1985 to 1995 quality	798–1101
'No-GAP' to 1995 with GAP	832
1985 to 'full GAP'	3854–4021
'No-GAP' to 'full' GAP	3986
1995 actual to 'full GAP'	2752–3189

In the analysis reported in the following sections, we have taken the range of values given above, as applying to 1995–6. In future years the numbers are increased:

- by 2 per cent per annum to allow for population growth;
- by 4 per cent per annum with an income elasticity of 0.285, to allow for real income growth (the net impact is to raise values by 1.14 per cent annually). The income elasticity of 0.285 was estimated using the survey data that generated the non-user benefit estimates.

User benefits

The key table for user benefits is Table 6.10. The same three types of improvements are analysed as for non-user benefits. Table 12.3 sums up the main user benefits. As for user benefits, there is a range when comparing the 1985 quality with the 1995 quality because the model generating the estimates has more than one version. The comparison of the annual benefits of 'Full GAP' quality with the quality that would exist without GAP reveals that the estimate is a comparison of higher than the corresponding estimate when the 1985 quality is taken. The difference is between 5 and 30 per cent, depending again on which comparison is taken. User benefits are, of course, much smaller than non-user benefits (by at least two orders of magnitude). It is possible that not all the user benefits have been captured in the CVM approach. The issue of what exactly is included in these figures has to be addressed, as there may be some double counting with the health benefits. This is discussed below.

TABLE 12.3
Summary of User Benefits Utilized in Analysis

(Rs million 1995)

Changes in water quality	Estimated annual benefits in 1995
1985 to 1995 quality	2.4–4.7
'No-GAP' to 1995 with GAP	3.1
1985 to 'full GAP'	15.0–16.0
'No-GAP' to 'full' GAP	16.7
1995 actual to 'full GAP'	10.2–13.6

In the analysis reported in the following sections, the range of values given above is taken as applying to 1995–6. In future years the numbers are increased:

- by 2 per cent per annum to allow for population growth
- by 4 per cent per annum with an income elasticity of 0.2, to allow for real income growth (the net impact is to raise values by 0.8 per cent annually). The income elasticity of 0.2 was estimated using the survey data that generated the user benefit estimates.

Health benefits

In the discussion of health benefits in Chapter 7, three sets of figures were provided: saving in working days for the general population, saving in working days for sewage farm workers, and saving in cost of water treatment.

The saving in working days was the result of a comparison of morbidity rates in areas before the GAP and in comparable areas after the

GAP. The estimated benefits were Rs 73.57 million per million users. The estimated number of users was 12,62,500, making a total benefit of Rs 92.88 million. The main issue here is whether these benefits can be taken in addition to the user benefits reported above. If the CVM had done its job properly, these health benefits would have been included. But the user benefits figures are much smaller (only about 2 to 3 per cent of the health benefits). Hence, it may be that the respondents were not fully aware of the health benefits, in which case it is appropriate to include them. On the other hand, the data on health are not that strong, so a case could be made for not including them. The results with these health benefits included are presented in the Appendix but in the summary tables it is shown how the overall assessment would change if they are not included. When included, the estimates were increased annually by 2 per cent for population growth and 4 per cent for real income growth.

The saving in working days for sewage farm workers was found to result from reduced protozoal infestation. Estimated grossed up benefits were Rs 16.8 million, and these have been included. They are increased over time in the same way as health benefits.

For both of the above categories of benefits, a sophisticated estimation of benefits as a function of water quality is not attempted. Hence, it is assumed that these benefits apply to the improvement for GAP II only, with further water quality improvements not generating any further benefits.

Finally, there were the benefits of reduced treatment costs. Some doubts were expressed about these estimates in Chapter 7. In addition, it would be incorrect to include them as well as including the health and other costs. The reason for this is double counting; if these treatments were undertaken we would not have the health impacts. So, one can either measure the health impacts or the costs of mitigation, but not both. For a number of reasons it is better to take the damage cost savings, which is what is done.

Agricultural benefits

Agricultural benefits are discussed in Chapter 10. They arise because of the irrigation benefits (the GAP water has some fertilizer, which results in higher yields even after allowing for the increased application of conventional fertilizer in non-GAP farms); the savings in fertilizer (GAP-affected farms use less conventional fertilizer than non-GAP affected farms); and the value of the sludge provided from STPs. The benefits of each of these categories were:

Incremental benefits from irrigation:	Rs 90.61 million
Benefits in reduced fertilizer application:	Rs 6.62 million
Benefits of use of sludge	Rs 12.02 million

These values have been included in the benefits. No increase in value over time is assumed for these benefits; nor is any allowance made for further improvements as water quality improves over and above the GAP I level.

Other benefits

The benefits not quantified in money terms are biodiversity, fisheries, and toxic effects. Some positive benefits for biodiversity and fisheries are well documented in the report; excluding these means that the estimates given below are underestimates of net benefits. It is difficult to say by how much, but at least for some aspects of biodiversity the non-monetary benefits should be considered as substantial. Part of the non-user benefits estimated in Chapter 5 can be attributed to the bio-diversity of the Ganges. For toxic effects, the effects are mixed; there are some improvements and some costs as well. Further work is required before these impacts can be quantified.

FINANCIAL ESTIMATES OF NET BENEFITS

The main results of the estimates of the net financial benefits are summarized in Tables 12.4 and 12.5. Table 12.4 gives a breakdown of the benefits by category and Table 12.5 gives the estimate of internal rates of return (IRRs). Table 12.4 shows the benefits from the GAP as being mainly in the non-user category (from 80 per cent to nearly 100 per cent, depending on which quality change is being evaluated). This makes the analysis very sensitive to those values. In order to see what would happen if the non-user benefits were ignored, the IRR is reported in Table 12.5 with and without the non-user benefits. The benefits and costs are estimated upto the year 2020 AD. Although there will be benefits after that date, it is difficult to estimate them after allowing for the additional costs that will be incurred in servicing the larger populations. Benefits are also assumed to start in 1987, and build up linearly to the full values discussed above by 1996.

TABLE 12.4
Distribution of Benefits from the GAP

(Rs million 1995)

Changes in water quality	*Estimated present value benefits in 1995*			
	Non-user	User	Agriculture	Health
1985 to 1995 quality (Model I)	6871	29	575	813
1985 to 1995 quality (Model II)	4922	15	575	813
'No-GAP' to 1995 with GAP	5136	19	575	813
1985 to 'full GAP' (Model I)	22,657	92	575	813
1985 to 'full GAP' (Model II)	21,720	83	575	813
'No-GAP' to 'full' GAP	22,556	92	575	868
1995 actual to 'Full GAP' (Model II)	11,252	47	0	0
1995 actual to 'Full GAP' (Model II)	9714	36	0	0

TABLE 12.5
Internal Rates of Return for the GAP: Financial Analysis

Water quality change	*With health benefits*		*Without health benefits*	
	With non-user benefits	Without non-user benefits	With non-user benefits	Without non-user benefits
1985 to 1995 Quality	13.6–17.4	0.9–1.1	12.5–15.6	(neg.)
'No-GAP' to 1995 with GAP	14.6	1.0	13.0	(neg.)
1985 to 'full GAP'	37.3–38.8	(neg.)	33.6–34.9	(neg.)
'No-GAP' to 'full' GAP	38.8	(neg.)	19.7	(neg.)
1995 actual to 'full GAP'	65.0–73.3	(neg.)	58.5–65.7	(neg.)

(neg.) = negative

Finally the analysis of GAP I + II is assumed to generate benefits from the year 2000 AD, assuming that the works will be completed by that date.[2] From 1997 to 2000, benefits are built up linearly as in the case of GAP I mentioned in the previous paragraph.

[2]At the time the study was undertaken, Phase II of the GAP was expected to be completed by the year 2000. However, it appears that it may take longer.

The main conclusions to be drawn from the above tables are:

(a) The improvement from 1985 quality to present quality has a return of around 13 to 17 per cent if all benefits are included. Leaving out the health benefits (which may be double counted) the return goes down to 12 to 16 per cent. Leaving out the non-user benefits lowers the return to around 1 per cent (with health benefits) and to less than zero (without health benefits).

(b) Replacing 1985 quality with the quality that would exist in 1995 with no GAP yields results that are within the range obtained with the models that generate the '1985 quality to 1995 quality' changes.

(c) The improvement from 1985 quality to full GAP quality, when both phases of the GAP are completed and functioning, is much higher (34 to 38 per cent, irrespective of whether the health benefits are included or not). But it is more sensitive to the non-user benefits. Without them the return is negative. This reflects the high marginal WTP for improvements in Ganges water quality to bathing quality standards by non-users.

(d) Following on from that, the improvement in quality from GAP I to GAP II has a very high IRR, the additional costs being small compared to the additional non-user benefits. The IRRs are around 65 to 73 per cent, but again, this is almost entirely due to the non-user benefits, without these the IRR is negative.

ECONOMIC ESTIMATES OF NET BENEFITS

The main results of the estimates of the net financial benefits are summarized in Table 12.6.

The additional conclusions to be drawn from this table are:

(a) That the economic rate of return for the 1985 to 1995 quality change is about one to two percentage points lower than the financial rate of return. The same applies for the '1985 to full GAP' change.

(b) For the marginal net benefits of going from GAP I to GAP II, the economic return falls by 4 to 5 per cent compared to the financial rates.

TABLE 12.6
Internal Rates of Return for the GAP: Economic Analysis

Water quality change	With health benefits		Without health benefits	
	With non-user benefits	Without non-user benefits	With non-user benefits	Without non-user benefits
1985 to 1995 quality	11.9–15.4	(neg.)	10.7–13.8	(neg.)
'No-GAP' to 1995 with GAP	12.3	(neg.)	11.1	(neg.)
1985 to 'full GAP'	36.4–37.8	(neg.)	18.4–18.8	(neg.)
'No-GAP' to 'full' GAP	37.6	(neg.)	33.8	(neg.)
1995 actual to 'full GAP'	61.6–69.2	(neg.)	55.4–62.3	(neg.)

(neg.) = negative

Other than that, the main conclusions of the financial analysis also hold for the economic analysis.

INCOME DISTRIBUTION EFFECTS OF THE GAP

BENEFICIARIES OF THE GAP

The cleaning of a very important river like the Ganges in the Indian subcontinent can produce significant income distribution benefits in the Indian economy. Being a lifeline for almost a third of the Indian population, a cleaner Ganges provides benefits to the rich as well as the poor. However, from the point of view of equity, benefits that accrue to the poor assume greater importance than those to the rich. Therefore, the quantification of income distribution effects is important in the Social Cost Benefit Analysis of the GAP.

The beneficiary groups of the GAP can be identified as urban users, non-urban users, non-users, skilled labour, unskilled labour, farmers, fishermen, industrial units, and government. User benefits are partly for amenity and partly for health. As explained in Chapters 5 and 6, surveys of users and non-users of the Ganges show that there are significant incremental user and non-user benefits from the GAP. Furthermore, as explained in Chapter 7, there are significant additional health benefits to urban and non-urban populations, not captured in the user surveys, which focus only on urban populations and may not elicit health benefits. Table 12.7 provides the estimates of household average annual willingness to pay for the user and non-user benefits of the urban population in India. Investment and operational expenditures on the GAP projects create significant employment benefits to skilled and unskilled labour. Farmers benefit from the increased crop productivity and savings in the use of fertilizers due to effluent treatment by the GAP projects. Given that there is evidence of underemployment of unskilled labour in India, there can be substantial incremental benefits for unskilled labourers employed on the GAP projects. The increased fish supply from the cleaning of the Ganges can increase the incomes of fishermen. However, the water polluting industrial units in the Gangetic basin incur costs for pollution abatement, for it is mandatory for them to achieve the water pollution standards set down in environmental laws. The government (both central and state) incurs the capital and operational costs of the GAP projects.

TABLE 12.7

Estimates of Household Annual Average Willingness-to-pay for User and Non-user Benefits of GAP

(Rupees)

	Urban user benefits	*Non-user benefits*	*User health benefits*
Current river quality	143.89	126.15	73.7
Bathing standard quality	450.20	441.38	73.7

Note: The health benefits are per user, not per household. Assuming a household of 5 persons, the benefits would be Rs 367.8.

Source: Chapters 5, 6, and 7

Table 12.8 provides typical estimates of present values of benefits of various beneficiaries during the 33 years of the GAP which have been analysed above. The benefits are calculated for the financial analysis and for the benefits of current quality relative to 1985 quality. The analysis uses variant one of the model for user and non-user benefits— that is for the case where the econometric model is employed without a water quality variable to estimate the WTP. To keep the analysis manageable, all figures reported in this section are for that one case only. Other benefit cases exhibit similar trends.

Beneficiaries like farmers, fishermen, most residents close to the river, and skilled and unskilled labourers belonging to lower income groups in the Indian economy share a significant part of incremental incomes arising out of the GAP. Given the currently prevailing under-employment to these groups of people in the Indian economy, the employment created by the GAP projects can lead to a redistribution of income in their favour. Estimation of the social benefits of the GAP requires that a relatively higher social premium has to be imputed to the incremental incomes accruing to people belonging to lower income groups in relation to the premium imputed to the incomes accruing to people belonging to higher income classes. The next section briefly describes a method of estimation of social premia or income distribution weights for the incomes of beneficiaries of the GAP.

TABLE 12.8

Estimates of Present Value of Benefits of Various
Beneficiaries at Market Prices

	(Rupees million)
1. Users	29.11
2. Non-users	6871.03
3. Farmers	574.93
4. Heath benefits of river users	826.93
5. Fishermen	n.a.
6. Skilled labour	1703.84*
7. Unskilled labour	1919.42†
8. Industrial units	−1504.59‡
9. Government	−4569.32

Notes: * These may not be taken as incremental benefits to skilled labour due to the GAP if there is full employment of skilled labour in the Indian economy.

     †These are the incremental benefits to unskilled labour if the direct opportunity cost of unskilled labour is zero in the Indian economy.

     ‡The cost incurred by the industrial units in the Gangetic basin for the water pollution abatement is not actually part of the GAP cost.

     n.a.—not available.

ESTIMATES OF INCOME DISTRIBUTION WEIGHTS FOR THE INCOMES OF THE GAP BENEFICIARIES

People receiving user and non-user benefits from the GAP belong to different income classes. Tables 12.9 and 12.10 provide information about the distribution of urban household samples for the user and non-user surveys by per capita income classes. A good number of households surveyed have per capita income around the Indian average national per capita income of Rs 9321 at 1995–6 prices (Government of India, 1996). The survey results show that 105 of 817 non-user households surveyed, and 150 of 576 user households surveyed have per capita incomes of less than Rs 10,000 at 1995–6 prices. Also, other beneficiaries of the GAP like farmers, fishermen, and unskilled labour belong to some of the lower income groups in the Indian economy.

TABLE 12.9

Distribution of Households for User Survey by Per Capita Income

Lower limit	Upper limit	Frequency	Relative frequency
0	10,000	150	0.2604
10,000	20,000	140	0.2431
20,000	30,000	82	0.1424
30,000	40,000	75	0.1302
40,000	50,000	39	0.0677
50,000	60,000	16	0.0278
60,000	70,000	29	0.0503
70,000	80,000	11	0.0191
80,000	90,000	2	0.0035
90,000	100,000	12	0.0208
100,000	110,000	0	0
110,000	120,000	1	0.0017
120,000	130,000	2	0.0035
130,000	140,000	2	0.0035
140,000	150,000	3	0.0052
150,000	160,000	0	0
160,000	170,000	1	0.0017
170,000	180,000	0	0
180,000	190,000	0	0
190,000	200,000	3	0.0052

TABLE 12.10
Distribution of Households for Non-user Survey
by Per Capita Income

Lower limit	Upper limit	Frequency	Relative frequency
0	10,000	105	0.1285
10,000	20,000	182	0.2228
20,000	30,000	139	0.1701
30,000	40,000	125	0.153
40,000	50,000	66	0.0808
50,000	60,000	38	0.0465
60,000	70,000	42	0.0514
70,000	80,000	25	0.0306
80,000	90,000	11	0.0135
90,000	100,000	26	0.0318
100,000	110,000	0	0
110,000	120,000	12	0.0147
120,000	130,000	4	0.0049
130,000	140,000	1	0.0012
140,000	150,000	11	0.0135
150,000	160,000	0	0
160,000	170,000	4	0.0049
170,000	180,000	0	0
180,000	190,000	1	0.0012
1,90,000	200,000	6	0.0073

Let us assume the following social welfare function for estimating the income distribution weights attributable to incomes of individuals belonging to different income groups in the Indian economy. More specifically, a special form can be taken for the social welfare function, and a common one that has been adopted is that of Atkinson (1970). He assumes that social welfare is given by the function

$$W = \sum_{i=1}^{N} \frac{AY_i^{1-\varepsilon}}{1-\varepsilon}, \qquad (12.1)$$

where
W = social welfare function
Y_i = income of individual i
ε = elasticity of social marginal utility of income or inequality aversion parameter
A = a constant

The social marginal utility of income is defined as

$$\frac{\partial W}{\partial Y_i} = AY_i^{-\varepsilon}. \tag{12.2}$$

Taking per capita national income, $\overline{Y}$ as the numeraire, and giving it a value of one, we have

$$\frac{\partial W}{\partial Y_i} = A\overline{Y}_i^{-\varepsilon} = 1, \tag{12.3}$$

and

$$\frac{\partial W}{\partial Y_i} = \text{SMU}_i = \left[\frac{\overline{Y}}{Y_i}\right]^{\varepsilon}, \tag{12.4}$$

where SMU_i is the social marginal utility of income going to a person in income group i relative to income going to a person with the average per capita income in India. The values of SMU_i are, in fact, the weights to be attached to costs and benefits to group i relative to costs and benefits to the person with income equal to the national per capita income.

In order to apply the method, estimates of $\overline{Y}_i$ and ε are required. Some recent studies on the estimates of inequality aversion parameter, ε, for the Indian economy (Murty *et al.*, 1992) provide estimates in the range of 1.75 to 2.0. Some other work for other countries has suggested a value closer to 1.0. The *Economic Survey, 1995–6* (Government of India), provides an estimate of per capita Gross Domestic Product for the Indian economy of Rs 9321 at 1995–6 prices.

Using this estimate of $\overline{Y}$ and values of ε, of 1.0, 1.75, and 2.0, the estimated income distribution weights attributable to different income classes in the Indian economy are given in Table 12.11.

ESTIMATES OF SOCIAL BENEFITS OF THE GAP

We assume that farmers (most GAP farmers are small and marginal farmers as explained in Chapter 10), health beneficiaries, unskilled labour, and fishermen belong to the Rs 5000 per capita income group in the Indian economy. Also, the income distribution preferences of the Indian government may be such that it is indifferent between giving a rupee at the margin to a person with an income equivalent to the per capita income in the Indian economy and keeping the rupee itself. This implies that the income distribution weight attributable to the income of the government is one.

TABLE 12.11
Income Distribution Weights

(Average Income is Rs 9321 per annum)

Income Class	Income Distribution Weights		
Rs 1995 p.a.	$\varepsilon = 1$	$\varepsilon = 1.75$	$\varepsilon = 2$
3000	3.107	7.271	9.653
4000	2.330	4.395	5.430
5000	1.864	2.974	3.475
6000	1.554	2.162	2.413
7000	1.332	1.651	1.773
8000	1.165	1.307	1.358
9000	1.036	1.063	1.073
10,000	0.932	0.884	0.869
11,000	0.847	0.748	0.718
12,000	0.777	0.643	0.603
13,000	0.717	0.559	0.514
14,000	0.666	0.491	0.443
15,000	0.621	0.435	0.386
20,000	0.466	0.263	0.217
25,000	0.373	0.178	0.139
30,000	0.311	0.129	0.097
35,000	0.266	0.099	0.071
40,000	0.233	0.078	0.054
45,000	0.207	0.064	0.043
50,000	0.186	0.053	0.035
55,000	0.169	0.045	0.029
60,000	0.155	0.038	0.024
65,000	0.143	0.033	0.021
70,000	0.133	0.029	0.018
75,000	0.124	0.026	0.015
85,000	0.110	0.021	0.012
95,000	0.098	0.017	0.010
105,000	0.089	0.014	0.008
115,000	0.081	0.012	0.007
125,000	0.075	0.011	0.006
135,000	0.069	0.009	0.005
145,000	0.064	0.008	0.004
155,000	0.060	0.007	0.004
165,000	0.056	0.007	0.003
175,000	0.053	0.006	0.003
185,000	0.050	0.005	0.003

p.a.—per annum

Table 12.12 gives estimates of income distribution weights for different beneficiary groups of the GAP. Given an estimate of inequality aversion parameter, ε, of 1.75 and governmental income as numeraire, a rupee of income accruing to unskilled labour, farmers, and fishermen has a social value of Rs 2.974, while the benefits accruing to users and non-users of the Ganges have social marginal utilities of Rs 0.096 and Rs 0.064 respectively. Very low social marginal valuation of benefits accruing to users and non-users of GAP is expected because their per capita incomes are 4 to 5 times higher than the national per capita income.

TABLE 12.12

Estimates of Income Distribution Weights for the Incomes of the GAP Beneficiaries

Beneficiary group	Annual per capita income (Rs)	Income distribution weight		
		$\varepsilon = 1$	$\varepsilon = 1.75$	$\varepsilon = 2.0$
Unskilled labour, farmers, and health beneficiaries	5000	1.864	2.974	3.475
Urban users	35,660	0.262	0.096	0.068
Non-users	44,705	0.209	0.064	0.043
Government	9321*	1.000	1.000	1.000

*National per capita income
Source: Estimated as explained in the text.

Table 12.13 gives the estimates of the net present value of the social benefits to the various GAP beneficiaries at a 10 per cent social rate of discount. The table shows that, allowing for the distribution effects increases the net benefits of the project, in spite of the fact that the gains of the major beneficiaries (non-users) are heavily discounted. The discounting effect is more than made up by the benefits of the GAP in benefits to low income groups (farmers, low income users' health benefits, and the employment benefits of unskilled labour).

Given that not all benefits accruing to lower income groups are included in this exercise, especially benefits from fish production, the social benefits of the project may be considered to be substantially in excess of the benefits estimated using market prices.

TABLE 12.13

Estimates of Social Benefits of the GAP (Current Quality Relative to 1985 Quality)

(Rupees million)

		$\varepsilon = 0.0$	$\varepsilon = 1$	$\varepsilon = 1.75$	$\varepsilon = 2.0$
1.	Users	29.11	7.53	2.79	1.98
2.	Non-users	6871.03	1417.57	439.74	295.45
3.	Health gainers	826.93	1515.70	2549.29	2873.58
4.	Farmers	574.93	1071.78	1709.84	1997.88
5.	Fishermen	n.a.	n.a.	n.a.	n.a.
6.	Unskilled labour	1919.42	3578.18	5708.36	6670.00
7.	Government	−4569.32	−4569.32	−4569.32	−4569.32
8.	NPV	5652.10	3021.44	5840.70	7269.57

n.a:—not available

Note: The cost to industry is not accounted.

FINANCIAL SUSTAINABILITY OF THE GANGA ACTION PLAN

ARE THE BENEFITS OF THE GAP ENOUGH TO JUSTIFY THE COSTS?

Studies aimed at the measurement of the benefits of the GAP in Chapters 5 and 6 show that there are significant user and non-user benefits from cleaning the Ganges. Also the cost–benefit analysis of the GAP attempted in the earlier sections of this chapter establishes that, by and large, the benefits from the GAP justify the costs incurred. Having recognized that the continuation of the GAP will benefit Indian society, one has to establish the source of funds for its financial sustainability. Adequate funds are required for the operation and maintenance of physical assets created during the GAP Phase-I, and for investment expenditures and the operation and maintenance of assets during the GAP Phase II.

The benefits from the GAP accrue to the entire Indian society and it may not be feasible to 'capture', that is to obtain payment in money for all the benefits generated by the government and other agencies spending on the GAP. However, it is necessary to have instruments and institutions to capture part of the benefits of the GAP to the Indian public, at least to the extent of recovering the capital and operating costs of the GAP projects. These institutions may be either government institutions or non-governmental organizations (NGOs). Household responses on the question of the payment vehicle in the user and non-user surveys described in Chapters 5 and 6 indicate that a majority of households do not have faith in the government to make the payments for the benefits they receive from the Ganges. However, a good number of surveyed households said that they did not mind making payments to an NGO or any philanthropic organization which can take responsibility for cleaning the Ganges. Any attempt to design institutions to make the GAP financially sustainable has to take cognizance of this public response to the question of payment vehicle in the contingent valuation surveys.

Various options can be considered for designing institutions and instruments for ensuring the financial viability of GAP projects. These can be based on any one of the following approaches:

(a) the 'polluter pays' principle;
(b) the 'user pays' principle;
(c) voluntary payments to NGOs; and
(d) provisions in the annual budget of the government.

Details of each of these approaches are discussed in the following sections.

THE POLLUTER PAYS PRINCIPLE

The main sources of pollution of the Ganges are industrial and household effluents, with household effluents contributing the major share. Under the 'polluter pays' principle the instruments used to recover the cost of the GAP should therefore be taxes levied or fees charged to the polluters. Environmental laws to achieve water pollution abatement by the water polluting industries in the Gangetic basin can be implemented by the use of either economic instruments like pollution taxes and marketable pollution permits, or command and control methods.

Under these regulations, industries are induced to install effluent treatment plants to reduce pollution loads to meet the environmental standards. In India, economic instruments have not been used so far to deal with the problem of environmental pollution. Instead, use of command and control measures by the government has led many water polluting industries in the Gangetic basin to build effluent treatment plants. Out of the 68 heavily water polluting industries in the Gangetic basin, 56 industries have such plants, while 12 industries were asked to close due to non-compliance with environmental regulations.

Serious attempts are now being made to design and implement a pollution tax aimed at industrial water pollution abatement based on the 'polluter pays' principle. Once such a tax is implemented, the inducement to control industrial pollution in the Gangetic basin at source will lessen the burden of treatment of industrial effluents on the GAP projects. Although non-complying industrial units will still be contributing effluents to be treated by the GAP projects, they will mainly have to deal with effluents from non-industrial sources, especially household effluents.

Pollution taxes on households and industries can raise revenue for financing the GAP projects. These taxes can be specially designed and levied on a 'per unit of effluent' basis by the government (Central or State Pollution Control Boards) or they can be collected as part of a tariff for household and industrial water use by the local water suppliers. The responsibility of designing and levying these taxes can be with the central government or Central Pollution Control Board while the State Pollution Control Boards or local governments take the responsibility for collecting them. Alternatively, they can be designed and collected by the State Pollution Control Boards taking into account the local ambient quality of the river. The net revenue collected after deductions for the operation and maintenance cost of the GAP projects can be distributed between the central and state governments in proportion to their shares of the capital cost of GAP.

A flat rate of tax on a kilolitre (KL) of effluent water can be collected from the households and industries in the Gangetic basin. The rate of tax can be computed, given the total annual cost of the GAP projects (annualized capital cost plus operation and maintenance cost) and the volume of household and industrial effluents. The annual cost of the GAP projects is estimated as Rs 1145.5 million at 1995–6 prices. The total annual volume of household effluents and untreated industrial effluents is estimated as 328.9 million kilolitre in the Gangetic basin (Table 12.14). Therefore, the cost of treatment of a kilolitre of effluent by the GAP projects is estimated as Rs 3.48. If the tax is levied at a flat rate of Rs 3.48 per kilolitre of effluent water on the polluters, the cost of the GAP can be recovered. Alternatively, an effluent fee or tax of this type can be collected as part of a water tariff by the water suppliers.

There may be some problems in the implementation of these taxes in India. First, as already mentioned, there are no pollution taxes at present in India. Second, the water tariffs are highly subsidized, so that any increase in these tariffs to collect revenue for financing the GAP projects could face political resistance. Third, the local governments or water suppliers may not have the incentive to collect such taxes, since they have to hand over net tax proceeds either to the state government or the central government.

TABLE 12.14

Quantity of Industrial and Household Effluents in the Gangetic Basin

(Million Kl/year)

Source of Effluents	Quantity of Effluents
Total annual household effluents intercepted by GAP projects	318.645
Total quantity of industrial effluents discharged by 68 gross polluting units along the Ganga	94.900
Quantity being treated by industrial units	84.680
Untreated industrial effluents	10.220
Total industrial and household effluents to be treated by GAP projects	328.865

Source: Government of India, Ministry of Environment & Forests, *Statistical Review of Programmes Under National River Conservation Directorate,* September, 1996.

THE USER PAYS PRINCIPLE

The user and non-user surveys conducted to elicit valuation of cleaning the Ganges by the urban households in India reveals that there is considerable demand for the environmental benefits offered by such a plan. The estimated annual willingness to pay by an Indian urban non-user household is Rs 449.06 while that of a user household is Rs 488.31 for achieving bathing quality standards. The respective values of willingness to pay for the current quality of the Ganges are Rs 274.79 and Rs 73.95. Therefore, it may be desirable to design some fiscal instruments to capture part of these benefits for the financial sustainability of the GAP projects.

For example, an additional income tax for cleaning the Ganges can be levied on all income tax payers in India. However, this instrument may not cover all the beneficiaries from the clean-up of the Ganges, since many of these may not be income tax payers. Therefore, there is an in-built progressiveness in such a tax, since income tax payers belong to relatively higher income groups in comparison to non-tax paying beneficiaries. Alternatively, in order to have a tax base covering all the beneficiaries from the clean-up of the Ganges, there can be a special environmental tax on both user and non-user households, with a built in progressiveness with respect to income of the households.

There are problems with the implementation of such tax instruments. First of all, there can be general objections to raising income tax. Secondly, the cost of assessing the incomes of all Ganges households, just for the sake of levying the special Ganges tax could be significantly high. One way to avoid these problems is to levy a flat rate of special Ganges tax on all the beneficiary urban households, such that the annualized capital cost, and the operation and maintenance cost of the GAP projects is recovered. Chapters 5 and 6 provide estimates of urban user and non-user beneficiary households of the Ganges respectively, of 87.33 million and 0.037 million, making a total of 87.4 million households approximately. Given the annual cost of the GAP projects of Rs 1145.5 million, the share of the total annual cost of the GAP is estimated as Rs 14.2 per beneficiary household, which is only 8 per cent of the average household willingness to pay of Rs 174.37 for the improved quality of Ganges due to the GAP.

VOLUNTARY PAYMENTS BY THE BENEFICIARIES

In many developing countries, the quality of the government creates problems for realizing the payments for the supply of public good services like the environmental benefits offered by the cleaned-up Ganges. As already pointed out, households surveyed using the contingent valuation method have refused to make payments to the government for cleaning the Ganges. They said, however, that they will be happy to make payments to any philanthropic organization or NGO which takes responsibility for cleaning the Ganges. Therefore, the creation of a NGO with governmental support to accept voluntary contributions from the beneficiaries of the Ganges may guarantee the financial sustainability of the GAP. While in the cases discussed earlier, the quality of the government is important because it has to play a major role, in this case, the government has to play only a catalytic or enabling role which is minimal. The capacity of such a fund to raise the amount required to finance the whole of the GAP remains uncertain. Some recent evidence on how much people actually pay compared to what they state they will pay in CVM studies indicates that a good part can be raised if the payment vehicle is properly designed. In any event, a combination of voluntary payments and other means could be the answer to the funding problems of the GAP.

The NGO created for mobilizing funds for the GAP should involve people belonging to different regions in the country and different political parties for its greater acceptability by the public. This can be done by having regional offices in the Gangetic basin and in different state capitals. It can accept contributions from domestic as well as foreign nationals for the clean-up of the Ganges.

GOVERNMENT BUDGETARY PROVISION FOR THE GAP

The annual cost of the GAP can form part of central and state budgets. In this case, the sharing of the cost of GAP is widely diffused among the taxpayers in India. However, there is a danger of the GAP not getting the required priority in the government's budget, given the priority of demands for other public good services like health, education, and sanitation in India.

CONCLUSIONS

This chapter has evaluated the net benefits of the two phases of the GAP. The costs of each phase are taken from Chapter 4 and the monetary benefits are taken from Chapter 5 (non-user benefits), Chapter 6 (user benefits), Chapter 7 (health), and Chapter 10 (agriculture). It is acknowledged that the benefits of biodiversity and fisheries could not be quantified in monetary terms. The impacts of toxic changes are less clear, but they too have not been quantified. Hence, the analysis is partial to the extent that these have been ignored.

Three kinds of changes have been analysed: the net benefits of: (a) going from 1985 water quality to 1995 quality, (b) going from 1985 quality to quality that will exist when GAP I and GAP II have been completed, and (c) the benefits of going from the present quality to the quality when GAP I and GAP II have been completed. In addition, we have looked at the impacts of making the comparison in terms, not of the 1985 quality, but the quality that would have prevailed in 1995 and thereafter, had GAP not been implemented. Because of higher pollution loading over time, the quality would have been worse in 1995 than it was in 1985 in the absence of the GAP.

The benefits are unevenly distributed across categories, with non-user benefits dominating. They account for anything from 80 per cent of the total to nearly 100 per cent of the total, depending on the comparison being made. In view of this, a check is also made of how much the IRR is affected by excluding the non-user benefits.

The financial IRR for GAP I is around 13 to 17 per cent, with the economic IRR being about one to two percentage points lower. The financial IRR for GAP I and GAP II taken together is around 33 to 39 per cent, with the economic IRR again being about one to two percentage points lower. The IRR for the change from GAP I to GAP II is much higher; around 65 to 73 per cent for the financial analysis and 61 to 69 per cent for the economic analysis. Without the non-user benefits, these IRRs fall to very low values and become negative in most cases. Moreover, the impact of non-user benefits is greater in the marginal valuation (GAP I to GAP II). In this case, the benefits are almost exclusively from non-users. It is also important to note that all the benefits assume that GAP II will achieve the goal of bathing water quality. Should it fail to do so, these benefits will not materialize.

The analysis of income distributional effects of the GAP was also attempted. The method employed has been to weight gains and costs according to who bears them. The distribution weights are estimated using a methodology developed by project appraisal economists. Applying these weights to the GAP results in some sharp changes in the allocation of benefits across groups. Without distribution weights, the main beneficiaries are the non-users. With such weights, the main beneficiaries are: unskilled labour employed by the project, those gaining from health improvements in the poor communities along the river, and the farmers. With these weights, the total benefits increase and the project generates a similar NPV in the case when the inequality aversion parameter takes the value of one, and a higher NPV (2 to 3 times higher) in the case when it takes the value of 1.75 to 2.0. The IRRs adjust in a corresponding way.

The important issue of how to finance the GAP on a sustainable basis is also dealt with in this chapter. A number of different mechanisms are suggested: a 'polluter pays' principle, a 'user pays' principle (with government involvement), a 'user pays' principle (without government involvement), and funding from the general tax system.

The 'polluter pays' principle would mean a water charge of Rs 3.48 per kilolitre, collected as a tax, or more likely, as part of the water tariff. This would be levied on all industrial effluents. The impact on industry and employment has not been part of the scope of this study, but is something that should be looked at. Given the general interest in employing market based instruments, this would make a good case study and first tax to be implemented. Problems to be resolved are how to ensure that the tax is collected efficiently and passed on to the authority with the responsibility for the implementation of the GAP.

The 'user pays' principle with government involvement places a tax on beneficiaries. A tax that does not distinguish between users and non-users is proposed. The identification of beneficiary households is a problem, as is the issue of how to get the tax increase approved through the political process.

An alternative to the involvement of the government is to set up a charitable commission for the Ganges with responsibility to collect payments from user and non-user beneficiaries. The commission would have autonomy from the government, but would be accountable to the public through the charities commission or a similar body. It would have regional offices, which would not only collect money, but would also be responsible for disseminating information about the Ganges and changes in its water quality. We believe that a good part of the necessary funding of the GAP could be raised in this way.

Finally, there is the possibility of using general taxes. We do not favour this approach, given the extreme pressures on such tax revenues and the uncertainties created for future revenue flows to maintain the river water quality.

Of course, there is no reason why only one of these options has to be implemented. We would favour, for example, a combination of an effluent tax and a voluntary payment scheme of the kind outlined above. The details have, of course, to be worked out, but the prospects of success are good.

13
Conclusions

SCOPE OF STUDY

This study was undertaken to review the Ganga Action Plan; in particular to make an appraisal of its costs and benefits. Such an exercise is unusual for river system clean-up programmes. Initially, the programme was carried forward on a tide of goodwill and strong public support. The question of comparing costs and benefits did not arise when the late Prime Ministers Indira and Rajiv Gandhi gave their support to the project. The importance of this study, however, was underscored by the fact that the costs were clearly substantial and questions were being raised as to whether (a) the money had been well spent, and (b) a country at the level of development, and with the limited public resources, of India, could successfully undertake such a programme. Success in this context is not only a question of finding the resources for the investments; it also means finding a way to finance the continuing operations of the programme on a sustainable basis.

COMPARISON WITH OTHER STUDIES

The present study was undertaken with this in mind. To start with, a discussion of similar programmes in other countries was attempted with a view that the GAP can be benefited from the experiences of these programmes. A number of important points that can be useful for the GAP emerged in the examination of the histories of clean-up programmes for the Thames, the Rhine, and the Danube in Europe.

First of all, all these programmes have a long history; actions have been taken over a period of 20 years or more. Therefore, it would be unreasonable for a developing country like India to expect to achieve in 10 years what it has taken countries with many more resources over 20 years to achieve.

Second, the costs involved in other river cleaning-up programmes have been enormous, and given the similar type of objectives with relatively larger-scale operations, the cost earmarked for the GAP is much smaller. The Thames investments have amounted to over £100 million over the period 1950–80, which works out to over Rs 5000 million at the 1995 exchange rates in India. The Rhine programme was even larger. For a 1320 kilometre long river, the expenditures have been around DM 100 billion, between 1965 and 1989, which amount to about Rs 1940 billion at the 1995 exchange rates. Finally, the Danube programme, for a river of similar length to the Ganges, a prioritized strategic action plan has been drawn up with an expected cost of $4 billion, or around Rs 125 billion. By contrast, the Ganga project, which covers over 2500 kilometres, has an investment budget for GAP I and GAP II of Rs 11,200 million. In this respect the GAP level of expenditure does not seem excessive.

Third, one may argue that the preservation of the quality of rivers is regarded politically as the objective of all civilized nations. Since whatever costs required are considered acceptable, there is no need for the cost–benefit analysis of the river cleaning-up programmes. But since the costs of river cleaning-up programmes are enormous and there are resource constraints in developing countries like India, the investments in river cleaning have to compete with investments in other development projects. Therefore, there is a need for the economic appraisal of river cleaning-up projects along with the other investment projects.

Fourth, doubts are often expressed about the sustainability of river cleaning-up programmes. The analysis of benefits and costs of these programmes will provide inputs for designing the policy instruments for sustainability. Only in the case of Danube Clean-up Programme has such a study been initiated recently. This study is therefore a timely contribution to the literature on the problem of measurement of benefits from the river cleaning-up projects.

METHODOLOGY

This study is an example of using the methodology of standard cost–benefit analysis for estimating the social benefits of an investment project for the preservation of environmental resources. Suitable extensions of standard methodology are made to deal with the special problems that arise in the valuation of benefits from the environmental resource, the Gangas, in the present case.

It highlights the difficulties involved in the identification and measurement of benefits and costs of investment projects for environmental
preservation. It describes the data requirement from primary and secondary sources. The type of analysis attempted in this study requires the
secondary data to be collected from different agencies in the economy
and the experience reveals that the required data is not properly organized by the concerned agencies in India. Given that in future there
may be many more environmental preservation projects like the GAP
in India, this study can provide guidelines for production and organization of the data for the agencies in both public and private sectors
related to environmental preservation.

This study also discusses and uses some of the special survey
methods for the collection of primary data to measure benefits from the
preservation of environment. Therefore, it can be a useful guide for the
estimation of environmental benefits from the water resources, along
with other similar studies in India and abroad.

COST OF CLEANING-UP

The cost of cleaning-up of the Ganges has to be borne by the
households and industries in the Gangetic basin according to the 'polluter pays' principle. The cost of the GAP is met by the Indian government (central government and the governments of Uttar Pradesh, Bihar,
and West Bengal). In addition to this cost, as shown in this study, water
polluting industries in the Gangetic basin are incurring a significant
amount of cost for cleaning-up the Ganges in the form of investment
and operating costs of water pollution abatement. The estimates of
benefits to households from cleaning-up the Ganges made in this study
can provide information to the government for designing appropriate instruments to charge the households for polluting the Ganges.

●

BENEFITS FROM CLEANING-UP

As shown in this study, the cleaning-up of the Ganges provides multiple
benefits. They are benefits accruing to people who stay near the river
or visit the river for pilgrimages or tourism. These will be in the form
of recreation and health benefits and are called user benefits. The other
category of benefits comprise those accruing to the people who are not
staying near the river but enjoy benefits by knowing that the river is
clean. This category of people can be both Indians and foreigners.

These are called non-user benefits arising out of people's preferences for the biodiversity or the aquatic life the Ganges supports and the religious significance of the river. It is also shown in this study that the GAP provides some types of irrigation and fertilizer benefits to farmers, and employment to skilled and unskilled labour. Also, the GAP contributes benefits in the form of cost savings to water supply undertakings along the Ganges. The estimates of these different types of benefits are obtained in this study.

The user and non-user benefits of cleaning the Ganges are estimated using the contingent valuation method (CVM) of survey of households. These benefits are described as people's willingness-to-pay (WTP) for cleaning the river. The CVM consisted of a carefully designed set of questionnaires to survey sample households in major cities in India, including the cities along the Ganges. This survey has evoked unexpectedly good responses about the households' valuation of user and non-user benefits for an important water resource in the country and provides a useful guide for similar future studies in India and abroad. The survey to estimate user benefits is aimed at estimating the benefits to the population living within one kilometre on either side of the river and to pilgrims. Therefore, the estimates of user benefits made may be understating the user benefits to the extent that a much larger population staying in the Gangetic basin is getting user benefits. The survey to estimate non-user benefits is aimed at the urban literate population in India. Given that there can be people having non-user benefits from cleaning the Ganges in rural households and illiterate people, especially because of the religious significance of the Ganges, there can be an underestimation of non-user benefits in this study.

The results of the non-user survey show that the average annual willingness-to-pay (WTP) for bathing quality water in the Ganga is Rs 558 per household per annum. This yields an aggregate figure of Rs 4873 million per annum for the 8.733 million households from the urban literate population in 23 major cities. The average annual WTP for current (1995) quality in the Ganga is Rs 193 per household per annum, which aggregates to Rs 1684 million per annum. The value of non-user benefits for past (1985) quality, that is the quality that existed before the GAP was initiated, is Rs 101 per annum per household, and totals Rs 852 million per annum.

The results of the user survey show that the estimate of WTP for user benefits from bathing quality water is Rs 581.59 per household per annum. It aggregates to Rs 19.019 million per annum for the 37,213 households in the target population of literate urban residents in the major cities in the states of Uttar Pradesh, Bihar, and West Bengal, who live within 0.5 kilometres of the river. Willingness-to-pay for benefits from current (1995) water quality in the Ganga is Rs 167.23 per household per annum, which totals Rs 5.5 million per annum. Similarly, household WTP for the direct-user benefits for the water quality existing in 1985 (i.e. before the GAP started) is Rs 93.28 per annum, which sums up to Rs 3.1 million per annum.

Reviewers to the draft report have noted that the WTP figures from the user and non-user surveys are similar (around Rs 500 for bathing water quality and around Rs 150 for current water quality). One might expect higher values for users than for non-users and one possible explanation offered was that the average incomes of literate households along the Ganga are lower than those of households living elsewhere (since Uttar Pradesh and Bihar are poor states). Thus, users may be willing to pay a higher percentage of incomes even if the absolute amounts are similar to other places. The average per capita income of user households is Rs 27,771 compared to Rs 36,467 for non-users, which is consistent with the above observation. Furthermore, it has been suggested that there may have been questionnaire design problems that have resulted in the similar figures. In this context, it is interesting to note that the values obtained for user benefits are not dissimilar to other studies. For the Gomti, for example, a recent DFID study reported a WTP of Rs 100–200 for user benefits for a cleaner river. This suggests that the bias referred to above is likely to be limited.

A comparison of these WTP estimates with those from other developing countries where the benefits of improvements in water quality levels have been assessed shows that the values obtained in this contingent valuation are broadly comparable.

HEALTH BENEFITS

This study has estimated the benefits of an improvement in water quality as a result of the GAP as Rs 77.57 per river user per annum. Assuming a household of five, this would amount to Rs 367.8. Earlier studies on health benefits from water quality improvement (Brandon and Hommann, 1995; Verma and Srivastava, 1990) have made some estimates but they are not attributable to particular kinds of water quality improvement; they do not focus particularly on river water quality. In particular the costs of water-borne diseases estimated by Verma and Srivastava come out at around Rs 915 per household per annum.

There are health benefits for sewerage farm workers from cleaning-up the Ganges. This study estimates them as Rs 4800 per worker per year. Assuming that there are 100 workers for each of the 35 GAP STP schemes, this would amount to a total benefit of Rs 16.8 million.

TOXIC IMPACTS

This study has examined the effects of changes in treated/untreated waste water toxicants (metals and pesticides) discharged by the STPs under the GAP on public health, agriculture, and environmental quality in the disposal (receiving) areas.

The effect of waste water toxicants (metals and pesticides) on human health in the areas receiving the effluents from STPs of the GAP was assessed through three different methods. They are (a) a standard questionnaire based survey of the exposed and unexposed population groups, (b) environmental exposure risk analysis, and (c) biomonitoring of the metals and pesticides levels in the human blood and urine of the different population groups under study areas. All the differnet approaches indicated a considerable risk and impact of heavy metals and pesticides on human health in the exposed areas receiving the waste water from STPs.

The impact of these treated waste water toxicants (metals and pesticides) on the environmental quality of the disposal areas was assessed in terms of their elevated levels in different media samples: water, soil, crops, vegetation, foodgrains, and biological samples collected from exposed areas, in comparison to the respective unexposed areas. The data generated shows their elevated levels in all the relevant environmental parameters, indicating a definite adverse impact on the environmental quality of the disposal areas.

The critical levels of the heavy metals in the soil for agricultural crops are found to be much higher than those observed in the study areas irrigated with effluents of the GAP projects. Therefore, there seems to be no adverse impact from metals and pesticides on agriculture in these areas. The sludge from the GAP projects has both positive and negative impacts on agriculture, as it is loaded with high levels of toxic heavy metals and pesticides, but also enriched with several useful ingredients such as N, P, K, etc. providing fertilizer values. The sludge studied had cadmium, chromium, and nickel levels above tolerable levels as prescribed for agricultural land application.

AGRICULTURAL BENEFITS

The agricultural benefits from the GAP projects come in three forms. First, there are the irrigation benefits arising from the partially treated water that is released to farms in the GAP area. The second benefit is from the fertilizer value of the irrigation water. It was observed that the GAP farms apply less fertilizer than the non-GAP farms. Finally, there are the benefits from the use of the sludge provided by the GAP projects as a fertilizer. The estimates of all these three forms of benefits add up to Rs 109.3 million per year.

FISHERIES

Fisheries have been an important source of benefits from cleaning the Ganges and continue to be so. An anlaysis of data related to the fish catch in the river is required to estimate the change in that catch that could be attributed to improved water quality. This could not be done in this study since the data required for such a sophisticated exercise were not available. Not only are the time series very short, they are in-complete for some stations and the corresponding data on effort are also incomplete. Hence, only a partial quantitative analysis of changes in the fisheries situation could be attempted in this study. An analysis of avail-able data is attempted to assess trends in catch at various stations on the Ganges, trends in catch per unit of effort of the fishermen, and a relationship between trends in catch and the improvement of water quality in the river.

BIODIVERSITY

The study has also assessed the impact of the GAP on the biodiversity of the Ganga ecosystem. Three areas were considered for this purpose. They are (a) species of international conservation significance (b) species that have commercial significance, and (c) overall assessment of ecosystem health.

There are a number of international species comprising mammals, reptiles, and birds supported by the Ganges ecosystem. The benefits from the GAP in preserving these species arise in four ways. First, there are some species for which there have been *in situ* conservation and captive breeding programmes. Second, the GAP has raised awareness and encouraged conservation efforts through information dissemination, etc. Third, the GAP has facilitated the collection of information on species and their habitat—something which will contribute in an important way to their conservation. Finally, the general improvement of the water of the Ganges has helped most of the above species.

The commercial species under biodiversity include otters, turtles, and crocodiles that are killed for commercial purposes. In general, the GAP has had little impact on these species, which are endangered because of land-use change and over-harvesting.

The overall ecosystem health is assessed in terms of specific biological parameters, namely the primary productivity, phytoplankton, zooplankton, benthic macroinvertebrates, and migratory fish. All these factors point to a situation that is slightly improved with respect to the early 1980s. Nevertheless, long term trends on ecosystem health are influenced by many factors, including pesticide loading, riverine forest cover, and hydraulic structures that change river morphometry, increase bank erosion, and create barriers for migratory fish. These trends persist, increasingly damaging the ecosystem.

NET BENEFITS OF THE GAP

The net benefits of the two phases of the GAP are estimated. The benefits of biodiversity and fisheries could not be quantified in monetary terms. The impacts of toxic changes were less clear and they too have not been quantified. Hence, the analysis is partial to the extent that these have been ignored.

The benefits are unevenly distributed across categories, with non-user benefits dominating. They account for anything from 80 per cent of the total to nearly 100 per cent of the total, depending on the comparison being made. In view of this, we have checked how much the IRR is affected by excluding the non-user benefits.

The financial IRR for GAP I and GAP II taken together is around 37–9 per cent. With the economic IRR again being about one percentage point lower without the non-user benefits, these IRRs fall to very low values. It is also important to note that all the benefits assume that GAP II will achieve the goal of bathing water quality. Should it fail to do so, these benefits will not materialize.

DISTRIBUTION IMPACTS OF THE GAP

The cleaning-up of a very important river like the Ganges in the Indian subcontinent could produce significant income distribution benefits in the Indian economy. Being a lifeline for almost a third of the Indian population, a cleaner Ganges provides benefits to the rich as well as the poor.

The beneficiaries of GAP can be identified as urban users, non-urban users, skilled labour, unskilled labour, farmers, fishermen, industrial units, and the government. These beneficiaries belong to different income groups in the Indian economy, and the incremental income to each group due to the GAP has to be weighted by the income distribution weight relevant to the income of that group. Applying these weights to the GAP results in some sharp changes in the allocation of benefits across groups. Without distribution weights, the main beneficiaries are the non-users. With the weights, their benefits become very small and those of users and unskilled labourers increase sharply. With these weights the total benefits increase and the IRRs adjust correspondingly. Given that not all benefits accruing to lower income groups are included in this exercise, especially benefits from fish production, the social benefits of the project may be considered to be substantially in excess of the benefits estimated using market prices.

SUSTAINABILITY

This study also discusses the important issue of how to finance the GAP on a sustainable basis. A number of different mechanisms were considered. They are a 'polluter pays' principle, a 'user pays' principle (with government involvement), a 'user pays' principle (without government involvement), and funding from the general tax system.

The 'polluter pays' principle would mean a water charge of Rs 3.48 per kilolitre collected as a tax, or more likely, as part of the water tariff. This would be levied on all industrial effluents. Given the general interest in employing market based instruments, this would make a good case study and would assist in the design of a suitable tax system. A particular problem to be resolved would be to ensure that the tax is collected efficiently and passed on to the authority with responsbility for the implementation of the GAP.

The 'user pays' principle with government involvement places a tax on beneficiaries. We have proposed a tax that does not distinguish between users and non-users, amounting to Rs 13.11 per beneficiary, which is only about 7.5 per cent of their stated WTP.

An alternative to the involvement of the government is to set up a charitable commission for the Ganges with the responsibility to collect payments from user and non-user beneficiaries. The commission would have autonomy from the government but would be accountable to the public through the Commission for Charities or a similar body. It would have regional offices, which would not only collect money, but would also be responsible for disseminating information about the Ganges and changes in its water quality. The recommendation of this study is that a good part of the necessary funding of the GAP could be raised in this way.

Finally, there is the possibility of using general taxes. This study does not recommend this approach, given the extreme pressures on such tax revenues and the uncertainties created for future revenue flows to maintain the Ganges water quality.

Of course, there is no reason why only one of these options has to be implemented. This study in fact favours, for example, a combination of an effluent tax and a voluntary payment scheme of the kind outlined above. The details have, of course, to be worked out, but the prospects of success are good.

FINAL VIEW ON THE STUDY

This study has demonstrated that cost–benefit analysis can be applied with some success to river action programmes. The results are interesting and relevant to policy makers in India. First, how one views the GAP depends on how one treats the distribution question. Without distribution weights, the benefits are largely to non-users, and they are large enough to justify the project. With distribution weights, the project is even more 'justifiable', but now it is on the grounds of the benefits it provides to low income users and on the benefits of income to unskilled workers employed in the construction sector.

Second, the marginal benefits of the project are much higher than the benefits of GAP I. If GAP II can deliver the water quality goals it is aiming for, the rates of return will be very high. Finally, there are practical and sensible ways of making the programme financially sustainable; this study has presented some thoughts on this, although further work is clearly needed.

Appendices

Appendix 5.1

WELFARE MEASURES OF NON-USER BENEFITS FROM THE GAP

This Contingent Valuation Survey provides data about the willingness-to-pay by non-users for the three scenarios of the quality of the Ganges: (1) Quality in 1985 (before the GAP), (2) Current Quality (1995), and (3) Best or Bathing Quality standard, the ultimate GAP objective. Each scenario corresponds to a certain level of supply of the public good, the water quality in the Ganga, for which households express their valuation. Since the quality of the Ganga is an increasing function of the GAP expenditures over time since 1985, the households' valuation may be expected to increase with the quality of the river. In that case, the households' utility function, U, can be defined as an increasing function of the public good, the quality of river water, Z, and the consumption of private goods represented by X $(x_1, x_2, \ldots, x_n)$

$$U = U(X, Z). \qquad (A.5.1.1)$$

If P is $n \times 1$ price vector of private goods and R is the price charged for the public good Z, the household budget constraint can be written as

$$P * X + R * Z = M, \qquad (A.5.1.2)$$

where M is income. Maximizing (A.5.1.1) subject to (A.5.1.2), demand functions for private goods can be derived as

$$x_i = x_i(P, M - R * Z, Z). \qquad (A.5.1.3)$$

The indirect utility function can be written from (A.5.1.1), (A.5.1.2) and (A.5.1.3) as

$$V = V(P, M - R * Z, Z). \qquad (A.5.1.4)$$

The expenditure function can be obtained by inverting (A.5.1.4) as follows

$$E^* = M - R, Z = E^*(P, Z, U). \qquad (A.5.1.5)$$

Minimizing (A.5.1.5) for the given U^0, the restricted expenditure function can be obtained as follows

$$E = E(P, R, Z, U^0). \qquad (A.5.1.6)$$

Now a welfare measure of a change in Z, the river quality, can be defined as

$$W_Z = \frac{\delta E}{\delta Z} \qquad (A.5.1.7)$$

which can be interpreted as the households' marginal willingness-to-pay for the improvement in river quality. Using the expenditure function in (A.5.1.6), the compensating and equivalent surplus measures of welfare change due to increase in Z in the three scenarios relative to the base scenario where river can not support any life system ($Z = Z^0$) can be defined as follows:

Compensating Surplus (CS):

$$CS = E(P, R, Z^0, U^0) - E(P, R, Z^1, U^0) = M - E(P, R, Z^1, U^0); \quad (A.5.1.8)$$

Equivalent Surplus (ES):

$$ES = E(P, R, Z^0, U^1) - E(P, R, Z^1, U^1), \qquad (A.5.1.9)$$

where U^0 and U^1 correspond respectively to individual utilities with base level of Z and the level of Z under consideration.

Household willingness to pay for the non-user benefits of river quality in a given scenario can be defined using the compensating surplus measure of welfare change as follows

$$WTP_h = M - E(P, R, Z^1, U^0, S_h) + \varepsilon_h^0 - \varepsilon_h^1, \qquad (A.5.1.10)$$

Alternatively, using the equivalent surplus measure equation (10) can be written as

$$WTP_h = E(P, R, Z^0, U^1, S_h) - M + \varepsilon_h^1 - \varepsilon_h^0, \qquad (A.5.1.11)$$

where S_h is a vector of socioeconomic characteristics of the hth household; and e_h^1 and e_h^0 are error terms pertaining to expenditure functions of the hth household.

Therefore, *WTP* can be written as a random variable which is a function of income and other household characteristics:

$$WTP_h = \alpha X_h + e_h, \qquad (A.5.1.12)$$

where X is a vector of explanatory variables, α is a vector of parameters, and e_h is multivariate normal, with a mean of 0 and a variance of μ^2.

Alternatively, since the household survey elicits the value of non-user benefits of households for the three levels of water quality in the Ganga, the willingness-to-pay function can be estimated by considering the river quality Z as an additional explanatory variable:

$$WTP_h = \alpha X_h + \beta Z + e_h \qquad \text{(A.5.1.13)}$$

Appendix 5.2

DERIVATION OF AN INDEX OF RIVER WATER QUALITY

Data on Biological Oxygen Demand (BOD) levels from the National Rivers Conservation Directorate (NRCD) of the Ministry of Environment and Forests, New Delhi, and from the simulations on the Water Quality Model done by the Industrial Toxicology Research Centre (ITRC), Lucknow have been used to construct a water quality index for the river for four different scenarios:

(1) Best or Bathing Quality
(2) Current Quality (1995)
(3) Past Quality (1985) and
(4) Simulated Current Quality (1995)

Essentially, data on BOD levels from 14 monitoring stations between Rishikesh and Rajmahal, common to both the data from the NRCD and the ITRC simulations, were used to create the Index for each of the four scenarios mentioned above. For each scenario, a weighted average of BOD levels was computed and inverted, since increasing BOD levels implies *lower* water quality.

The weights were calculated as a ratio of incremental distance to total distance where, for a reading of BOD level from a particular monitoring station, the incremental distance was taken as the distance between that monitoring station and the next, and the total distance was the distance between the first monitoring station and the last. For example, in the case of the Kannauj Upstream (U/S) monitoring station, the incremental distance is the distance between Kannauj U/S and Kannauj D/S, that is 3 kilometres, and the total distance is the distance between Rishikesh and Rajmahal, 1508 kilometres. The value of the weight is thus 3/1508.

This weighted average was finally multiplied by 100 so that the Index for Best or Bathing Quality was 100, and the others calibrated accordingly to 48.63 in the case of Current (1995) Quality and 31.64 for Past (1985) Quality.

Best Quality was taken to be 1 milligram of BOD per litre, although the BOD level commensurate with Bathing Quality is defined as being less than 3 milligrams per litre, and that with Drinking Quality as *less than 1 milligram per litre*. This was done for two reasons. First, since BOD is only one of many parameters that define Bathing Quality (or Drinking Quality), setting BOD levels for bathing quality less than 3 milligrams per litre was considered an appropriate measure to take into account the other factors that define water quality (such as the level of dissolved oxygen (DO), the faecal coliform bacteria count, low concentrations of heavy metals and other toxic substances, pH, etc.).

Second, for Bathing Quality defined by values of BOD higher than 1 milligram per litre (for example 1.5 or 2.0), the Index value for Current Quality (1995) turned out to be nearly the same as Bathing Quality. Given that it is widely believed that this is *not* the case currently, the best option was to calculate the Index for Best Quality with BOD levels set at 1 milligram per litre uniformly.

Appendix A

MATHEMATICAL MODELLING OF THE GANGA RIVER WATER QUALITY

INTRODUCTION

One of the tasks of this study has been to estimate the Ganga river quality during 1995, both with and without the Ganga Action Plan Phase I. The continued Ganga river quality monitoring programme since 1986 has been carrying out river quality measurement/assessment at 27 selected locations only. From the analytical data at these sites, it may be erroneous to extrapolate the levels/trends for water quality parameters as a continuous function. Also, for generating the river water quality scenario in 1995, in the absence of the Ganga Action Plan (Phase I), there is no alternative other than a modelling approach. It was necessary, therefore, to follow the Water Quality Modelling Approach. This component on the water quality modelling of the Ganga river is basically to develop the river water quality scenarios (with particular reference to the Dissolved Oxygen and Biochemical Oxygen Demand) for the year 1995, both with and without the Ganga Action Plan.

WATER QUALITY MODEL INPUT DATA

RIVER HYDRAULICS AND HYDROLOGICAL DATA

As the standard model input data set on the Ganga river/major tributaries, their hydraulics and hydrological data (river width, mean depth, area of cross-section, flow velocity, Manning's Coefficient, discharge, etc.) at various locations between Rishikesh (UP) and Rajmahal for the year 1995 were collected from different regional and field offices of the Central Water Commission (CWC), Ministry of Water Resources, Government of India with the help of the National River Conservation Directorate (NRCD), Ministry of Environment and Forests, Governement of India.

RIVER WATER QUALITY DATA

For the purpose of calibration of the Water Quality Model, the Ganga river/tributaries quality data (temperature, Dissolved Oxygen, Biochemical Oxygen Demand, etc.) for 1995 (January to June 1995) were obtained from the NRCD, New Delhi. Sampling stations are marked on Figure 8.1 in Chapter 8.

RIVER LOADING WITH URBAN WASTEWATER/INDUSTRIAL EFFLUENTS

For different urban settlements/industrial units in the river basin, their waste water/industrial effluents generated during 1995 were obtained from the NRCD. Also, the wastewater/effluents fractions as intercepted/diverted/treated under the Ganga Action Plan strategies until 1995 were obtained from NRCD.

MODEL CALIBRATION

For the Ganga river water quality modelling, the total river stretch between Rishikesh and Rajmahal (1520 kilometers) was divided into 11 different reaches. The Water Quality Model (TOMCAT) was first calibrated with the help of river hydraulics/hydrological and water quality data, as obtained for 1995. The model output for DO and BOD was compared with the measured values available for selected locations. The model calibration output, both for river water DO and BOD are presented in Figure A.1 and A.2.

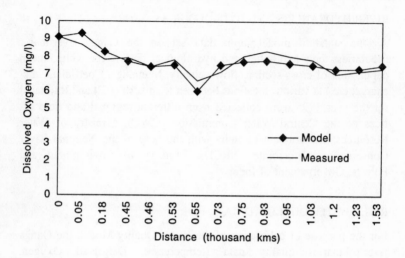

FIGURE A.1:   Model Calibration for Dissolved Oxygen in 1995 (GAP):
Model and Measured vs Distance

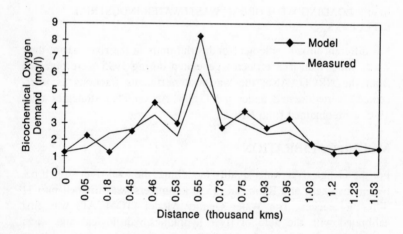

FIGURE A.2:   Model Calibration for Biochemical Oxygen Demand in 1995
(GAP): Model and Measured vs Distance

THE GANGA RIVER WATER QUALITY WITH THE GAP

The Water Quality Modelling (WQM) predictions for the DO and BOD levels in the Ganga river during 1995 with GAP Phase-I are presented in Figures A.3 and A.4. From the plots of river water DO and BOD (milligrams per litre) versus distance (Figures A.3 and A.4 and Table 4.1), it can be noted that after the GAP Phase-I, a total stretch of about 437 kilometres still violated the permissible level of 3.0 milligrams per litre for BOD. However, in the case of DO, throughout the river stretch, it was more than 5.0 milligrams per litre.

THE GANGA RIVER WATER QUALITY WITHOUT THE GAP

To generate the Ganga river water quality scenario with respect to DO and BOD, had there been no Ganga Action Plan, the estimated urban wastewater/industrial effluents (1995) from various towns/industries in the basin were taken as emptying completely into the Ganga river without any interception/diversion/treatment etc. The calibrated model was run to predict the DO and BOD levels for 1995. The output results are presented in Figures A.5 and A.6 and Table A.1. The plots of DO and BOD versus distance (river stretch, kilometres) show that the river stretch of about 740 kilometres (out of the total 1520 kilometres) between Rishikesh and Rajmahal would have violated the BOD limit of 3.0 milligrams per litre, and also the DO level could have been below 5.0 milligrams per litre in about 6.5 kilometres stretch. Further, a stretch of more than 100 kilometres would have witnessed BOD higher than 10.0 milligrams per litre.

WATER QUALITY MODELLING OF THE GANGA RIVER

The Ganga river *WQM* exercise developing scenarios in terms of the Dissolved Oxygen (DO) and Biochemical Oxygen Demand (BOD) levels of the river water during the year 1995, both with and without the Ganga Action Plan Phase-I, with a view to assessing the real impact of GAP output for scenarios indicates that a total river stretch of about 437 kilometres, out of a total stretch of 1520 kilometres between Rishikesh (UP) and Rajmahal (Bihar) still has a BOD level above the permissible limit of 3.0 milligrams per litre. The problematic stretch mainly lies between Kannauj and Varanasi. Further, had there been no GAP, a total river stretch of about 740 kilometres between Rishikesh

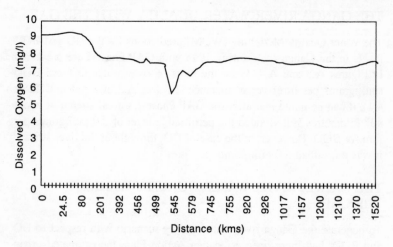

FIGURE A.3: Ganga River Water Quality in 1995 (with GAP):
Dissolved Oxygen vs Distance

and Rajmajal (which includes within it the 437 kilometres currently polluted) would have had BOD levels above 3.0 milligrams per litre and
of this, a more than 100 kilometres stretch could have exceeded BOD
levels of 10 milligrams per litre. DO levels of the river water would
have also been below 5.0 milligrams per litre in a few kilometres long
stretch near Kanpur.

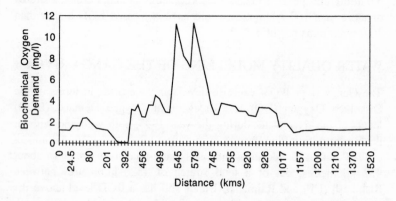

FIGURE A.4: Ganga River Water Quality in 1995 (with GAP):
Biochemical Oxygen Demand vs Distance

CONCLUSIONS ON WATER QUALITY IMPACTS OF THE GAP

The data on water quality is relevant to the valuation of the GAP because we need to know what the water quality would have been in the absence of the Action Plan. Table A.1 provides us with part of the information. DO and BOD levels have been computed for different locations in 1995, with and without the GAP. The 'without GAP' levels are

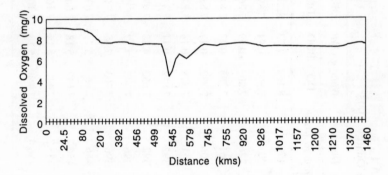

FIGURE A.5: Ganga River Water Quality in 1995 (No GAP):
Dissolved Oxygen vs Distance

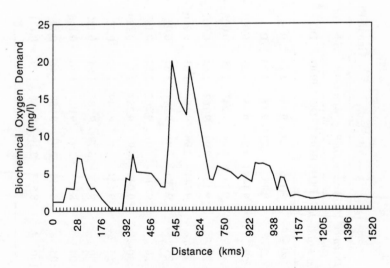

FIGURE A.6: Ganga River Water Quality in 1995 (No GAP):
Biochemical Oxygen Demand vs Distance

TABLE A.1
Ganga River Water Quality, 1995

Distance (km) from Rishkesh	Without GAP		With GAP		Effect of GAP on		Distance (km) from Rishkesh	Without GAP		With GAP		Effect of GAP on	
	DO	BOD	DO	BOD	DO	BOD		DO	BOD	DO	BOD	DO	BOD
0	9.136	1.24	9.136	1.24	0	0	176	8.168	1.73	8.274	1.18	0.106	-0.55
2	9.15	1.23	9.15	1.23	0	0	201	7.802	0.91	7.897	0.64	0.095	-0.27
2	9.148	1.28	9.15	1.25	0.002	-0.03	226	7.803	0.49	7.861	0.35	0.058	-0.14
20	9.24	1.24	9.243	1.21	0.003	-0.03	261	7.785	0.22	7.819	0.16	0.034	-0.06
20	9.16	2.99	9.162	1.64	0.002	-1.35	306	7.832	0.17	7.839	0.13	0.007	-0.04
24.5	9.144	2.96	9.16	1.63	0.016	-1.33	351	7.825	0.14	7.828	0.1	0.003	-0.04
28	9.275	2.91	9.297	1.61	0.022	-1.3	392	7.812	0.11	7.813	0.08	0.001	-0.03
28	9.077	7.07	9.262	2.34	0.185	-4.73	392	7.652	4.28	7.694	3.19	0.042	-1.09
30	9.098	6.97	9.287	2.31	0.189	-4.66	396	7.745	4.17	7.759	3.11	0.014	-1.06
55	9.129	5.1	9.349	1.92	0.22	-3.18	396	7.703	7.48	7.715	3.62	0.012	-3.86
80	9.001	3.79	9.183	1.64	0.182	-2.15	456	7.66	5.2	7.696	2.5	0.036	-2.7
104	8.865	2.95	9.013	1.46	0.148	-1.49	456.2	7.642	5.18	7.689	2.5	0.047	-2.68
105	8.848	2.91	8.988	1.46	0.14	-1.45	456.2	7.442	5.14	7.465	3.63	0.023	-1.51
140	8.537	2.19	8.658	1.29	0.121	-0.9	456.3	7.444	5.13	7.468	3.62	0.024	-1.51

Distance (km) from Rishkesh	Without GAP		With GAP		Effect of GAP on		Distance (km) from Rishkesh	Without GAP		With GAP		Effect of GAP on	
	DO	BOD	DO	BOD	DO	BOD		DO	BOD	DO	BOD	DO	BOD
456.3	7.756	4.95	7.766	4.47	0.01	−0.48	698.6	7.32	5.89	7.47	3.74	0.15	−2.15
466	7.41	4.64	7.434	4.19	0.024	−0.45	746.6	7.472	4.11	7.551	2.72	0.079	−1.39
499	7.417	3.87	7.434	3.49	0.017	−0.38	746.6	7.416	5.84	7.517	3.73	0.101	−2.11
531	7.505	3.24	7.517	2.93	0.012	−0.31	749.6	7.358	5.76	7.48	3.68	0.122	−2.08
532	7.484	3.21	7.498	2.9	0.014	−0.31	749.6	7.493	5.38	7.632	3.66	0.139	−1.72
532	7.28	8.65	7.37	6.33	0.09	−2.32	754.6	7.551	5.33	7.671	3.62	0.12	−1.71
539.6	6.312	20.04	6.689	11.18	0.377	−8.86	779.6	7.658	5.09	7.741	3.46	0.083	−1.63
544.6	4.473	17.32	5.703	9.66	1.23	−7.66	844.6	7.708	4.51	7.772	3.06	0.064	−1.45
548.6	4.85	14.84	5.996	8.26	1.146	−6.58	854.6	7.711	4.42	7.777	3	0.066	−1.42
555.6	6.082	13.91	6.746	7.77	0.664	−6.14	854.6	7.701	4.67	7.776	3.04	0.075	−1.63
563.6	6.515	12.92	7.007	7.25	0.492	−5.67	919.6	7.722	4.14	7.789	2.69	0.067	−1.45
563.6	6.305	19.08	6.86	11.35	0.555	−7.73	921.6	7.669	4.07	7.754	2.65	0.085	−1.42
578.6	6.106	16.62	6.736	9.92	0.63	−6.7	924.6	7.605	3.98	7.712	2.59	0.107	−1.39
598.6	6.453	13.85	6.953	8.31	0.5	−5.54	924.6	7.52	6.2	7.68	3.41	0.16	−2.79
623.6	6.797	11.08	7.162	6.71	0.365	−4.37	925.6	7.472	6.15	7.658	3.38	0.186	−2.77
648.6	7.034	8.91	7.304	5.46	0.27	−3.45	925.6	7.47	6.3	7.655	3.41	0.185	−2.89

Distance (km) from Rishkesh	Without GAP		With GAP		Effect of GAP on		Distance (km) from Rishkesh	Without GAP		With GAP		Effect of GAP on	
	DO	BOD	DO	BOD	DO	BOD		DO	BOD	DO	BOD	DO	BOD
927.6	7.385	6.2	7.616	3.35	0.231	-2.85	1204.6	7.257	1.6	7.295	1.2	0.038	-0.4
937.6	7.144	5.73	7.504	3.1	0.36	-2.63	1204.6	7.249	1.83	7.293	1.26	0.044	-0.57
937.6	7.26	4.71	7.52	2.81	0.26	-1.9	1205.6	7.253	1.83	7.297	1.26	0.044	-0.57
1016.6	7.248	2.74	7.499	1.64	0.251	-1.1	1205.6	7.253	1.98	7.297	1.27	0.044	-0.71
1016.6	7.246	4.36	7.497	1.83	0.251	-2.53	1209.6	7.271	1.97	7.315	1.27	0.044	-0.7
1019.6	7.223	4.27	7.497	1.79	0.274	-2.48	1209.6	7.249	1.86	7.285	1.29	0.036	-0.57
1069.6	7.26	2.96	7.432	1.4	0.172	-1.56	1215.6	7.241	1.85	7.277	1.29	0.036	-0.56
1155.6	7.363	1.93	7.465	1.08	0.102	-0.85	1252.6	7.237	1.84	7.276	1.28	0.039	-0.56
1156.6	7.377	1.93	7.477	1.08	0.01	-0.85	1295.6	7.289	1.82	7.325	1.28	0.036	-0.54
1156.6	7.355	1.84	7.432	1.19	0.077	-0.65	1369.6	7.375	1.76	7.413	1.26	0.038	-0.5
1160.6	7.358	1.82	7.431	1.18	0.073	-0.64	1369.6	7.374	1.78	7.412	1.28	0.038	-0.5
1160.6	7.357	1.84	7.431	1.21	0.074	-0.63	1411.6	7.468	1.75	7.501	1.27	0.033	-0.48
1169.6	7.368	1.8	7.433	1.19	0.065	-0.61	1459.6	7.55	1.74	7.581	1.26	0.031	-0.48
1169.6	7.437	1.69	7.488	1.23	0.051	-0.46	1459.6	7.617	1.78	7.643	1.38	0.026	-0.4
1199.6	7.229	1.6	7.269	1.2	0.04	-0.4	1519.6	7.489	1.73	7.511	1.36	0.022	-0.37

the ones used in one of the scenarios in Chapters 6 and 7 to compute the benefits of improved water quality.

The other scenarios take the actual 1985 quality and value the improvements relative to actual quality in 1995 and projected quality after GAP I and GAP II have been completed. We have not shown the actual water quality levels for 1985 in this chapter, but the data are available on request. Comparing the 1985 quality with the 1995 quality without GAP shows a deterioration of around 5 per cent in DO levels.

References

Abelson, P. (1996), 'Valuing Water Quality in Wuxi, China', in *Project Appraisal and the Environment: Valuation of the Environment*, London, ODI, St. Martins Press.

AIIH & PH (1997), *A Study on Health Impacts of the Ganga Action Plan*.

Anderson, J. (1879), *Anatomical and Zoological Researches: Comprising an Account of Zoological Results of the Two Expeditions to Western Yunnan in 1868 and 1875*; and *A Monograph of the Two Cetecean Genera Platanista and Orcaella*, London, B. Quaritch, two volumes.

Andrews, M.J. (1984), 'Thames Estuary: Pollution and Recovery', in P.J. Sheehan, D.R. Miller, G.C. Butler and P. Bourdeau (eds), *Effects of Pollutants at the Ecosystem Level*, London, John Wiley & Sons Ltd, pp. 195–227.

Anger, W. and Kent (1989), 'Human Neurobehavioural Toxicology Testing: Current Perspectives', *Toxicology and Industrial Health*, 5(2), pp. 165–80.

Atkinson, A.B. (1970), 'On the Measurement of Inequality', *Journal of Economic Theory*, 2(3), pp. 244–63.

Arrow, K., R. Solow, P.R. Portney, E.E. Leamer, R. Radner and H. Schuman, (1993), 'Report of the NOAA on Contingent Valuation', *Federal Register*, 58(10).

Baillie, J. and B. Groombridge (eds) (1996), *IUCN Red List of Threatened Animals*, The World Conservation Union, Switzerland.

Banerjee, Brojendra Nath (1989), *Can the Ganga be Saved?*, Delhi, B.R. Publishing Corporation, p. 197.

Basu, A.L. (1965), 'Observations on the Probable Effects of Pollution on the Primary Productivity of the Hooghly and Mutlah estuaries', *Hydrobiologia*, 25(1–2), pp. 302–17.

Bennett, J.W. (1984), 'Using Direct Questioning to Value Existence Benefits of Preserved Natural Areas', *Australian Journal of Agricultural Economics*, 28, pp. 136–52.

Benoist, A.P. and G.H. Broseliske (1994), 'Water Quality Prognosis and Cost Analysis of Pollution Abatement Measures in the Rhine Basin (The River Rhine Project: EVER)', *Water Science Technology*, 29(3), pp. 95–106.

—— (1979), 'Limnological Survey and Impact of Human Activities on the River Ganges', Technical Report, MAB Project-5 (UNESCO), pp. 1–91.

—— (1985), 'Ecology of the River Ganges: Impact of Human Activities', Final Technical Report, pp. 1–97.

Bilgrami, K.S. and J.S. Datta Munshi (1992), 'Bioconservation and Biomonitoring of River Ganga in Bihar', Technical Report submitted to the Ganga Project Directorate, Government of India, New Delhi.

Bowker, J.M. and J.R. Stoll (1988), 'Use of Dichotomous Choice Non-market Methods to Value the Whooping Crane Resource', *American Journal of Agricultural Economics*, 70, pp. 372–81.

Brandon, Carter, and Kirsten Hommann (1995), *The Cost of Inaction: Valuing the Economy-Wide Cost of Environmental Degradation in India*, Asia Environment Division, Washington, DC, World Bank.

Brookshire, D.S., L.S. Eubanks and A. Randall (1983), 'Estimating Option Prices and Existence Values for Wildlife Resources', *Land Economics*, 59, pp. 1–15

Carson, R. (1962), *Silent Spring*, Boston, Houghton Mifflin.

Central Inland Capture Fisheries Research Institute (ICAR) (1997), *Database on Fisheries of the Main Ganga*, Barrackpore, West Bengal.

Choe, K.A., D. Whittington and D.T. Lauria (1994), 'Household Demand for Surface Water Quality Improvements in the Philippines: A Case Study of Davao City', mimeo, Washington, DC, The Environment Department of World Bank.

Choudhury, B.C. and S. Bhupathy (1993), *Turtle Trade in India*, New Delhi, WWF–India, pp. 1–50.

CIFRI/ICAR (1985), *Annual Report*, Barrackpore, West Bengal, India, p. 133.

Das, I. (1991), *Colour Guide to the Turtles and Tortoises of the Indian Subcontinent*, Avon, England, R & A Publishing Limited, pp. 1–133.

Das Gupta, P.S., S.A. Marglin and A.K. Sen (1972), *Guidelines for Project Evaluation*, New York, United Nations.

Das Gupta, S.P. (1984), *Basin Sub-basin Inventory of Water Pollution: The Ganga Basin, Part II*, Central Board for the Prevention and Control of Water Pollution, New Delhi, ADSORBS/7/1982–83, p. 204.

Datta Munshi, J.S. (1992), 'Technical Report Submitted to the Ganga Project Directorate, London, Government of India', New Delhi Development Administration.

Diamond, P. (1996), 'Discussion of the Conceptual Underpinnings of the Contingent Valuation Method by A.C. Fisher' in D.J. Bjornstad and J.R. Kahn (eds), *The Contingent Valuation of Environmental Resources: Methodologi-*

cal Issues and Research Needs, Cheltenham, UK and Brookfield, US, Edward Elgar.

Ehrlich, P. and A. Ehrlich (1981), *Extinction: The Causes and Consequences of the Disappearance of Species*, New York, Random House, p. 38.

—— (1992), 'The Value of Biodiversity', *Ambio*, 21(3), pp. 219–26.

Esrey, S.A., J.B. Potash, L. Roberts and C. Shiff (1991), *Effects of Improved Water Supply and Sanitation on Ascariasis, Diarrhoea, Dracunculaisis, Hookworm Infection, Schistosomiasis, and Trachoma*, WHO, Geneva.

European Commission (1995), 'ExternE: Externalities of Energy', Vol. 2, *Methodology*, Brussels, European Commission, DG12.

European Commission DGXII (1995a), 'Externalities of Energy "ExternE"' Project, Science, Research and Development Programme (JOULE), Summary Report, EUR 16520 EN.

European Commission DGXII (1995b), Externalities of Energy 'ExternE' Project, Science, Research and Development Programme (JOULE), Methodology Report, EUR 16521 EN.

Fisher, A.C. (1996), 'The Conceptual Underpinnings of the Contingent Valuation Method' in D.J. Bjornstad and J.R. Kahn (eds), *The Contingent Valuation of Environmental Resources: Methodological Issues and Research Needs*, Cheltenham, UK and Brookfield, US, Edward Elgar.

Fitter, R. (1986), *Wildlife for Man: How and Why We Should Conserve Our Species,* IUCN/SSC Source Book for UNEP, pp. 1–223.

Foster-Turley, P., S. Macdonald and C. Mason (1990), *Otters: An Action Plan for Their Conservation*, IUCN/SSC Otter Specialist Group pp. 1–126

—— (1979), *The Benefits of Environmental Improvement: Theory and Practice*, Baltimore, The Johns Hopkins University Press for Resources for the Future.

Freeman, III, A.M. (1993), *The Measurement of Environmental and Resource Values: Theory and Methods*, Washington, DC, Resources for the Future.

Government of India (1996), *Economic Survey, 1995–96*, Ministry of Finance.

Gren, I., K. Groth and M. Sylven (1995), 'Economic Values of Danube Floodplains', *Journal of Environmental Management*, 45, pp. 333–45.

Grewal, B. (1995), *Birds of the Indian Subcontinent*, Hong Kong, The Guidebook Co Ltd, p. 1193.

Groombridge, B. (1982), *The IUCN Amphibia-Reptilia Red Data Book*, Switzerland, IUCN-Gland, p. 426.

Hanninen, H. and K. Lindstrom (1979), *Behavioural Test Battery for Toxicopsychological Studies Used at the Institute of Occupational Health in Helsinki*, Helsinki, Institute of Occupational Health.

Hochin, J.P. and A. Randall (1989), 'Too Many Proposals Pass the Benefit-cost Test', *American Economic Review*, 79, pp. 544–51.

Hora, S.L. (1952), 'Major Problems of the Fisheries of India, with Suggestions for Their Solution', *Journal of the Asiatic Society of Science*, 17(1), pp. 83–100.

Imhoff, K.R., P. Koppe, and E.A. Nusch (1991), 'Toxic Substances Management Principles Derived from Experience with Water Quality Management in the Ruhr River Basin', in *Guidelines in Lake Management: Toxic Substances Management in Lakes and Reservoirs*, 4, International Lake Environmental Committee Foundation/UNEP, Otsu, Japan, pp. 127–59.

Jhingran, A.G. (1991), 'Impact of Environmental Stress on Freshwater Fisheries Resources' *Journal of Inland Fish Society of India*, 23(2), pp. 20–32.

Jones, S. (1982), 'The Present Status of the Gangetic Susu, *Platanista gangetica* Roxburgh, with Comments on the Indus Susu, *P. minor* Owen', *FAO Fish Ser. (5) 4*, FAO Advisory Committee on Marine Resources Research Working Party on Marine Mammals, pp. 97–115.

Kannan. K., K. Senthilkumar and R.K. Sinha (1997), 'Sources and Accumulation of Butyltin Compounds in Ganga River Dolphin, *Platanista gangetica*', *Applied Organometalic Chemistry*, 11, pp. 223–30.

Kannan, K., R.K. Sinha, S. Tanabe, H. Ichihashi and R. Tatsukawa (1993), 'Heavy Metals and Organochlorine Residues in Ganges River Dolphin from India', *Marine Pollution Bulletin*, 26(3), pp. 159–62.

Kannan, K., S. Tanabe, R. Tatsukawa and R.K. Sinha (1994), 'Biodegradation Capacity and Residue/Pattern of Organochlorines in Ganges River Dolphins from India', *Toxicological and Environmental Chemistry*, 42, pp. 249–61.

Khan, M.A. (1996), 'Investigation on Biomonitoring and Ecorestoration Measures in Selected Stretches of the Rivers Ganga and Yamuna', Final Technical Report submitted to National River Conservation Directorate, New Delhi, Government of India, pp. 1–64.

Krishnamurti, C.R., K.S. Bilgrami, T.M. Das and R.P. Mathur (1991), *The Ganga: A Scientific Study*, New Delhi, Northern Book Centre.

Lakshminarayana, S.S. (1965), 'Studies on the Phytoplankton of the River Ganges, Varanasi, India, Part II: The Seasonal Growth and Succession of the Plankton Algae in the River Ganges', *Hydrobiologia*, 25(1 & 2), pp. 138–66.

Leatherwood, S. and W.A. Walker (1979), 'The Northern Right Whale Dolphin (*Lissodelphis borealis* Peale) in the Eastern North Pacific', in H.E. Winn and B.L. Olla (eds), *Behaviour of Marine Animals, Current Perspectives in Research.*, Vol. 3: Cetaceans, NY, Plenum Press, pp. 85–141.

Little, I.M.D. and J.A. Mirrlees (1974), *Project Appraisal and Planning for Developing Countries*, London, Heinemann Educational Book.

Luken, R. (1987), 'Economic Analysis: Canal Cities Water and Waste Water Phase II', unpublished report for USAID, Cairo.

Majid, I., J.A. Sinden and A. Randall (1989), 'Benefit Evaluation of Increments to Existing Systems of Public Facilities', *Land Economics*, 59, pp. 377–92.

Markandya, A. (1991), *The Economic Appraisal of Projects: The Environmental Dimension*, Washington, DC, Inter-American Development Bank, p. 115.

Mäler, K.G. (1974), *Environmental Economics: A Theoretical Enquiry*, Baltimore, The Johns Hopkins Press for Resources for the Future.

Manolis, S.C. (1987), *Wildlife Management: Crocodiles and Alligators*, Chipping Norton, Australia, Surrey Beatty and Sons.

Markandya, A., P. Harou, V. Cistulli and L. Bellu (1999), *Environmental Economics for Sustainable Growth: A Handbook for Policy-Makers*, Cheltenham, Edward Elgar.

Markandya, A. and D. Pearce (1991), 'Development, the Environment, and the Social Rate of Discount', *The World Bank Research Observer*, 6(2), pp. 137–52.

McConnell, K.E. and J.H. Ducci (1989), 'Valuing Environmental Quality in Developing Countries: Two Case Studies', Atlanta, Georgia, paper presented at the Allied Social Science Association.

Messel, H. and G.C. Vorlicek (1989), *Ecology of Crocodyles porosus in Northern Australia*. in N.S. Giland (1989), *Crocodiles: Their Ecology, Management and Conservation*, a Special Publication of the IUCN/SSC, Crocodiles Specialist Group, Switzerland, pp. 164–83.

Messel, H, F. King Wayne and J. P. Ross (1992), *Crocodiles: An Action Plan for Their Conservation*, IUCN/SSC.

Mitchell, R.C. and R.T. Carson (1981), 'An Experiment in Determining Willingness to Pay for National Water Quality Improvements', draft report to the US Environmental Protection Agency, Washington, DC.

—— (1984), *A Contingent Valuation Estimate of National Freshwater Benefits: Technical Report to the US Environmental Protection Agency*, Washington, DC, Resources for the Future.

—— (1989), *Using Surveys to Value Public Goods: The Contingent Valuation Method*, Washington, DC, Resources for the Future.

—— (1995), 'Current Issues in the Design, Administration and Analysis of Contingent Valuation Surveys' in P.O. Johansson, B. Kriström and K.G. Mäler (eds), *Current Issues in Environmental Economics*, Manchester and New York, Manchester University Press.

Murty, M.N. (1997), 'Integrating Environmental Considerations into Economic Policy Making: Institutional Arrangements and Mechanism at National Level in India', *ESCAP*, Bangkok.

Murty, M.N., B.N. Goldar, G.K. Kadekodi, and S.N. Mishra (1992), 'National Parameters for Investment Project Appraisal in India', Working Paper No. E/153/92, Delhi, Institute of Economic Growth.

―――― (1999), *Economics of Industrial Pollution: Indian Experience*, Delhi, Oxford University Press.

Myers, N. (1979), *The Sinking Ark*, New York, Pergamon.

―――― (1983), *A Wealth of Wild Species*, Boulder, Westview Press.

Natarajan, A.V. (1989), 'Environmental Impact of Ganga Basin Development on Gene-Pool and Fisheries of the Ganga River System', in D.P. Dodge (ed.), *Proceedings of the International Large River Symposium*, Can. Spec. Publ. Fish Aquat. Sci., 106, pp. 545–60.

National Environmental Engineering Research Institute (NEERI) (1995), *Annual Report*, Nagpur.

ODA (1990), 'Bogota Sewage Treatment Plant', unpublished report of the Overseas Development Administration, London.

ONL/RFF (1994), 'Estimating Fuel Cycle Externalities: Analytical Methods and Issues', Report 2, Oakridge National Laboratory and Resources for the Future, New York, McGraw-Hill.

Otto, D., I. Shalik, H.K. Hudnell, J. Rateliffe, and D. House (1994), 'Association of Mercury Exposure with Neurobehavioural Performance of Children in Bohemia', *Neurotoxicol.*, 15(4), p. 962.

Pahwa, D.V. (1979), 'Studies on the Distribution of the Benthic Macro Fauna in the Stretch of River Ganga', *Indian J. Anim. Sci.,* 49(3), 212–19.

Pahwa, D.V. and S.N. Mehrotra (1966), 'Observations on Fluctuations in the Abundance of Plankton in Relation to Certain Hydrological Conditions of River Ganga', *Proc. Nat. Acad. Sci. India*, 36(2), pp. 157–89.

Pearce, D., D. Whittington, S. Georgiou and D. Moran (1994), 'Economic Values and the Environment in the Developing World', a report to the United Nations Environment Programme, Centre for Social and Economic Research on the Global Environment (CSERGE), University College London and University of East Anglia, UK, and University of North Carolina at Chapel Hill.

Rai, L.C., (1978), Ecological Studies of Algal Communities of the Ganges River at Varanasi, *Ind. J. Ecol.*, 2(1), pp. 1–6.

Randall, A. (1989), 'Nonuse Benefits: Measuring the Demand for Environmental Improvement' in J.B. Braden and C.D. Kolstad (ed.), *ILENR/RE.ES 89/18.*, University of Illinois, Urbana, Sec. 10.1–10.19.

Rao, R.J. (1996), 'Studies on Biological Restoration on Ganga River in Uttar Pradesh: An Indicator Species Approach', Report submitted to National River Conservation Directorate, Government of India, New Delhi, pp. 1–181.

Rao, R.J., D. Basu, S.M. Hasan, B.B. Sharma, S. Molur and S. Walker (eds), (1995), 'Report on Population and Habitat Viability Assessment Workshop for Gharial', organized by Jiwaji University, Gwalior and Forest Department of Madhya Pradesh.

Ray, A. (1984), 'Cost Benefit Analysis: Issues and Methodologies', Baltimore, Johns Hopkins Press.

Raymond, N., L. Jane, S. Kenneth and S. Dennis (1995), 'Hypotheses to Explain the Higher Symptoms Rates Observed Around Hazardous Waste Sites', *SOZ Praventivemed*, 40(4), pp. 209–17.

Reinhard, E.G. (1931), 'The Plankton Ecology of the Upper Mississippi, Minneopolis to Winona', *Ecol. Monogr.*, 1, pp. 395–464.

Roy, H.K. (1955), 'Plankton Ecology of the River Hooghli at Palta', *Ecology*, 36(2), 164–75.

Royal Commission on Environmental Pollution (1972), HMSO, London.

Sharma, R.C. (1991), *Reptiles in Faunal Resources of Ganga, Part I*, Zoological Survey of India, New Delhi, pp. 27–50.

Shetty, H.P.C., 'Observations on the Distribution and Fluctuations of Plankton in the Hooghli–Matlah Estuarine System with Notes on their Relation to Commercial Fish Landings', *Indian Journal on Fish*, 8(2), pp. 326–63.

Sinha, R.K. (1996), 'Final Technical Report on Bioconservation of the Gangetic Dolphin, *Platanista gangetica*', submitted to the National River Conservation Directorate, Government of India, p. 1069.

Sinha, R.K. and K. Prasad (eds) (1988), *Final Technical Report of Ganga Basin Research Project*, Patna, Patna University, pp. 1–88.

Smith, M.A. (1931), *The Fauna of British India including Ceylon and Burma, Reptilia and Amphibia-Loricata*, London, Testudines-Taylor and Francies I.

Smith, V.K. (1987), 'Non Use Values in Benefit Cost Analysis', *Southern Economic Journal*, 54, pp. 19–26.

Smith, V.K. and J.V. Krutilla (1982), 'Towards Reformulating the Role of Natural Resources in Economie Models', in V.K. Smith and J.V. Krutilla (eds), *Explorations in Natural Resource Economics*, Baltimore, The Johns Hopkins University Press for Resources for the Future.

Squire, L. and H. van der Tak (1975), *Economic Analysis of Projects*, Baltimore, Johns Hopkins Press.

Stern, N., M.J. Srtis and A.R. Nobay (eds) (1977), 'The Marginal Valuation of Income', *Studies in Modern Economic Analysis,* Oxford, Blackwell.

Stoll, J.R. and L.A. Johnson (1984), 'Concepts of Value, Non-Market Valuation, and the Case of the Whooping Crane', Transactions of the Forty-ninth North American Wildlife and Natural Resources Conference, 49, pp. 182–393.

Tikadar, B.K. (1983), *Threatened Animals of India*, Calcutta, Zoological Survey of India, p. 285.

Tikadar, B.K. and R.C. Sharma (1985), Handbook of Indian Testudines, Zoological Survey of India, 12, p. 152.

Verma, B.L. and R.N. Srivastava (1990), 'Measurement of the Personal Cost of Illness Due to Some Major Water-related Diseases in an Indian Rural Population', *International Journal of Epidemiology*, 19(1), pp. 169–76.

Vorlicek, G.C. (1989), *Crocodiles: Their Ecology, Management and Conservation*, Switzerland, IUCN/SSC Crocodile Specialist Group, IUCN—The World Conservation Union, NS Gland.

Webb, G.J.W., P.J. Whitehead and S.C. Manolis (1987), 'Crocodile Management in the Northern Territory of Australia' in G.J.W. Webb, S.C. Manolis and P.J. Whitehead (eds), *Wildlife Management: Crocodiles and Alligators*, Chipping Norton, Australia, Surrey Beatty and Sons, pp. 107–24.

Weisbred, B.A. (1968), 'Income Distributional Effects and Benefit Cost Analysis' in S.B. Chase, Jr. (eds) *Problems in Public Expenditure Analysis*, Washington, DC, The Brookings Institution.

Whitaker, R. (1987), 'The Management of Crocodiles in India' in G.J.W. Webb, S.C. Manolis and P.J. Whitehead (eds), *Wildlife Management: Crocodiles and Alligators*, Chipping Norton, Australia, Surrey Beatty and Sons.

Whitaker, R. and Z. Whitaker (1989), 'Ecology of the Mugger Crocodile', *Crocodiles: Their Ecology, Management and Conservation*, Crocodile Specialist Group, IUCN—The World Conservation Union Publication, New Series, pp. 276–97.

Whitehead, H. and J.E. Carscaden (1985), 'Predicting Inshore Whale Abundance—Whales and Capelin off the Newfoundland Coast', *Can. J. Fish. Aquat. Sci.*, 42, pp. 976–81.

Williams, P.L. (1996), *Environmental and Occupational Health Effects of Pesticides and Other Chemicals*, SUDA/Cooperative State Res. Ser Fedrip Database, NITS.

Winpenny, J.T., (1991), *Values for the Environment: A Guide to Economic Appraisal*, London, Her Majesty's Stationery Office, Overseas Development Institute.

Wood, L. (1982), *The Restoration of the Tidal Thames*, Bristol, Adam Hilger, pp. 202.

World Bank (1994), *World Development Report, 1994.*

———— (1995), *Environmental Policy for the Middle East and North Africa*, Grey Cover, February 1995.

Index